Construction, Demolition and Disaster Waste Management

An Integrated and Sustainable Approach

T0174742

Construction, Demolition and Disaster Waste Management
An Integrated and Sustainable Approach

Erik K. Lauritzen

CRC Press
Taylor & Francis Group
Boca Raton London New York

CRC Press is an imprint of the
Taylor & Francis Group, an **informa** business

CRC Press
Taylor & Francis Group
6000 Broken Sound Parkway NW, Suite 300
Boca Raton, FL 33487-2742

First issued in paperback 2020

ISBN-13: 978-1-4987-6821-4 (hbk)
ISBN-13: 978-0-367-65711-6 (pbk)

Library of Congress Cataloging-in-Publication Data

Names: Lauritzen, Erik K., author.
Title: Construction, demolition and disaster waste management : an integrated and sustainable approach / Erik K. Lauritzen.
Description: Boca Raton : Taylor & Francis, a CRC title, part of the Taylor & Francis imprint, a member of the Taylor & Francis Group, the academic division of T&F Informa, plc, [2019] | Includes bibliographical references and index. |
Identifiers: LCCN 2018020248 (print) | LCCN 2018028305 (ebook) | ISBN 9781498768221 (Adobe PDF) | ISBN 9781351650076 (ePub) | ISBN 9781351640558 (Mobipocket) | ISBN 9781498768214 (hardback) | ISBN 9781315154022 (ebook)
Subjects: LCSH: Construction and demolition debris--Recycling. | Concrete--Recycling. | Masonry--Recycling. | Buildings--Salvaging.
Classification: LCC TD899.C5885 (ebook) | LCC TD899.C5885 L39 2019 (print) | DDC 628.4--dc23
LC record available at https://lccn.loc.gov/2018020248

Visit the Taylor & Francis Web site at
http://www.taylorandfrancis.com

and the CRC Press Web site at
http://www.crcpress.com

This book is dedicated to my wife, Mette Lauritzen,

thanks for your support and understanding of my absence.

Contents

Preface

The building and construction sector consumes a major part of the total resource consumption of the world. The sector produces one-third of all waste and is a main contributor to CO_2 emission. Referring to the agenda of the UN Sustainable Development Goals 2030, the Climate Agreement in Paris 2015 and the EU initiatives on Circular Economy, it is obvious to look at the resource management of the building and construction sector focusing on resource efficiency, reduction and recycling of waste materials.

Starting in the demolition blasting business in 1978, I realized that it was not the performance of demolition work alone that should be considered to complete a successful job. Control of the environmental impact and waste management was of equal importance. In 1981, I met Prof. Torben C. Hansen, who had just started the RILEM[1] technical committee TC 37-DRC on demolition and reuse of concrete. We began a long-term teamwork on demolition and recycling of concrete together with a number of international experts and fiery souls. Since the 1980s, the RILEM technical committees within the field of recycling have produced a great number of reports, proceedings, recommendations and guidelines on recycling. On this basis, standards for recycling crushed concrete and masonry, including the use of crushed concrete as aggregates in new concrete, were introduced in Denmark, The Netherlands and other countries – later as background for preparations for the EU CEN standard on the use of recycled aggregates in concrete. In the early 1990s, three dwelling houses were constructed in Denmark with 80% of recycled materials: concrete with crushed concrete aggregate, reused bricks and tiles, recycled windows, doors and floors. We were convinced that the high-quality recycling technologies were ready and just around the corner for implemention in the construction and building sector as commonly accepted technology. At the same time, the EU Commission

---

[1] RILEM: Réunion Internationale des Laboratoires d´Essais et de Recherches sur Matériaux et les Construction/International Union on Testing and Research Laboratories for Materials and Structures.

declared construction and demolition waste (CDW) as a priority waste stream, and lots of effort and money were allocated to research and development of recycling technologies and CDW management. However, the common use of recycled materials in the past years was limited to the use of crushed concrete in unbound materials for roads and fill. No serious progress in recycling CDW had taken place until the introduction of the Circular Economy concept in the mid-2010s. The demand by the construction sector for recycled materials, or secondary raw materials, does not match the interest of the waste sector for recycling CDW. Many barriers are mentioned, such as the risk of pollution, lack of technical quality and documentation, unreliability of supply and general difficulties in planning and management. In particular, I find the lack of information and education to be one of the most crucial barriers.

Inspired by the RILEM work, combined with EU-supported research and development projects as well as 40 years of practical experience on a number of demolition and recycling projects, I have found that sharing my observations and ideas on *integrated and sustainable approaches* has contributed to the mutual understanding of CDW and resource management. In addition, I found an opportunity to include disaster waste management, because the management of waste materials from damaged and destroyed buildings in principle is very much like the management of CDW. The principles of integrated approach are based on the general understanding and respect of all stakeholders' mutual interest in transformation of CDW to resources. The principle of a sustainable approach requires an understanding of the life cycle of buildings and structures – in particular, the stages and the time of life cycles. Compared with other products, buildings have a long life cycle, typically 50 years or more. When it comes to end-of-life of buildings or materials, we must consider the opportunities of keeping the value of the assets through transformation of the life cycle from an (older) life cycle to another (new) life cycle. Therefore, the transformation processes, whether they are called recycling, up- or down-cycling, reuse, recovery or other names, must be kept in mind.

Natural disasters and conflicts cause damage to thousands of buildings and result in millions of tons of debris every year. Therefore, I find it appropriate to transfer the experiences from CDW management in the building and construction sector to the aid sector. Reconstruction with maximum use of recycled materials puts people to work and saves time, resources and money. It is my ambition that the organizations involved with the clearance and reconstruction of damaged buildings and structures after disasters and conflict will exploit the opportunities for recycling of debris and other building materials.

I wish to thank the many persons and organizations for their support of my work with this book. In 2015, Prof. Dr Enrico Vasquez, Spain, Chairman of RILEM Technical Committee 223-MSC 2005–012; Dr Mats Torring, Stena Recycling, Sweden and Martin Bjerregaard, 3D

Consulting and Disaster Waste Recovery, UK, reviewed my proposal for the book and paved the way to publishing. I am very grateful to Solvejg Qvist, NIRAS, Denmark, for allowing me to use the figures and project information from my work at the time with DEMEX (1978–2004) and NIRAS (2004–2013). The City Concept of the EU IRMA[2] project has been highlighted in the book with acknowledgement of the work of the IRMA partners: NIRAS DEMEX, Dansk Betonteknik A/S, Brandis A/S, SBS Byfornyelse, and Meldgaard A/S (Denmark), Intron BV, Rotterdam Public Works Engineering Consultants, and Delft Technical University (The Netherlands), Demoliciones Tecnicas S.A. (Spain), Belgian Research Institute, Enviro Challenge, and Brussels Institute for Management of Environment (Belgium), Contento Trade Srl (Italy), Dr Tech. Olav Olsen a.s. (Norway), Federal State of Bremen and Hochschule Bremen, Laboratory for Building Materials (Germany).

Based on many years of cooperation with Anders Henrichsen, Dansk BetonTeknik, project manager of the EU-supported IRMA project, I have learned about practical recycling of concrete. I am very happy about the cooperation with Peter Laugesen, Pelcon Materials and Testing, and his contribution to the book on *Concrete Technology and Recycling Concrete*. I very much appreciate that he has taken the time to review Section 4.2 "Recycling of Concrete".

Special thanks to the Danish Ministry of Foreign Affairs, the World Bank, UNDP, UNWRA, Disaster Waste Recovery, UK, and Engineers without Borders, Denmark, for the opportunity to introduce recycling after conflicts and disasters in Bosnia, Kosovo, Lebanon, Turkey, Haiti and Nepal. Thanks to the Institute of Demolition Engineers for the use of the logo of the institute. The Copenhagen Municipality has given recycling of CDW a high priority, and I thank Jonny Christensen for the permission to present the flagship experiences of recycling in Copenhagen. The book presents the transformation of the Carlsberg Breweries into Carlsberg City as a case story with an acknowledgement to Carlsberg for the ongoing efforts on the reuse of old buildings and reuse of masonry.

Presenting the future aspects of recycling, I thank 3XN, Danish architects for the permission of presenting their impressive rehabilitation project in Sidney. Thanks to Kasper Guldager Jensen and Casper Østergaard, 3XN, and John Sommer, MT Højgaard, Denmark, for figures and information on the design for disassembly of the Circle House project. Thanks to Jakob Jørgensen, 3D Printhuset, Denmark, for information on 3D printing and visits to the 3D-printing demonstration of the 3D project in Copenhagen.

---

[2] IRMA: Integrated Decontaminated and Rehabilitation of Buildings, Structures and Materials in Urban Renewal. European Commission Fifth Framework Programme Energy, Environment and Sustainable Development Key Action 4: City of Tomorrow and Cultural Heritage, Contract no. EVK4-CT-220-00092.

Thanks to all other persons and organizations for granting permission for the use of figures and material. I appreciate the inspiring cooperation of Jette Bjerre Hansen, Danish Competence Centre on Waste and Resources (DAKOFA) and the DAKOFA Network of CDW Management and thank DAKOFA for all their support. I thank Peter Munch Jørgensen in particular for drawing the figures and the preparation of the photos. I am very happy about his contribution to the book. Finally, I thank Gabriella Williams and the publisher, CRC Press/Taylor & Francis, for their cooperation and patience during the work with the book.

Erik K. Lauritzen
Lauritzen Advising
Denmark

Author

Erik K. Lauritzen is managing director and owner of Erik Krogh Lauritzen ApS, Lauritzen Advising. He has 40 years of experience in the fields of explosive engineering, demolition and recycling, risk and safety assessment, post-war reconstruction, disaster management and project management. He is especially involved, as past Chairman of the RILEM Technical Committee on Guidelines for Demolition of Concrete and Masonry, in research and development of demolition and rubble recycling technologies.

Lauritzen has been responsible for a number of major demolition projects and post-disaster and post-war reconstruction projects in Lebanon, Kosovo, Bosnia, Turkey, Japan and Haiti in cooperation with the World Bank, UNDP, UNRWA, UNEP, EU and DANIDA. From 2012 to 2018, he was chairman of the Danish Network for Construction and Demolition Waste under the Danish Centre of Competence on Waste and Resources (DAKOFA).

Construction, Demolition and Disaster Waste Management

Construction and Demolition Waste (CDW), from the construction, maintenance, renovation and demolition of buildings and structures, represents a large proportion of the waste in industrialized societies.

Compared to other forms, such as household waste, more than 90% of CDW can be used as a resource and a substitute for construction materials, especially for primary and natural raw materials. Reuse, recovery and recycling depends on the quality and market for the materials, and the environmental impact of the processes for conversion of CDW from old structures to its use in new structures. However, the utilization today of CDW products as secondary resources is marginal. Most CDW is deposited or used as fill material, and the opportunities for high-quality recycling are generally neglected.

This book presents the opportunities for the sustainable and resource-efficient utilisation of CDW, focusing on recycling of concrete and masonry as the major forms of CDW. The recycling of gypsum, timber, mineral wool, asphalt and other types are also described. Its aim is to present a chain of value and material streams in the transformation of obsolete buildings and structures into new buildings and structures. It takes a holistic view, focusing on the life-cycle economy (the circular economy) and integrated management aspects of various scenarios ranging from high industrial urban renewal to debris removal and management after disasters and conflicts.

It is based on the author's 40 years of research and development combined with practical international experience within the demolition and recycling arena. It addresses students, architects, civil engineers, building owners, public authorities and others working in urban planning, demolition and resource management in the building and construction sector and in the reconstruction of damaged buildings after disasters and wars.

Chapter 1

Introduction

Life is understood to be backward but must be lived forward.

Søren Kierkegaard, Danish Philosopher and Author, 1813–1855

1.1 SUSTAINABLE BUILDINGS AND MATERIALS

1.1.1 Shortage of resources

Since the oil crisis in the 1970s, the industrial countries have recognized the need for natural resources because of the coherence between economic activity and the state of the environment has undergone a growing understanding. The environment (nature) provides various resources, services and goods, for example air, water, earth and light for our activities. However, the consumption and the rate of utilisation of the resources might exceed the capacity of the earth. There is a certain limit to non-renewable resources.

In 1972, *The Limits to Growth* by Dennis L. Medows (Medows 1972) warned about overconsumption of oil and gas and a number of minerals. Some years later, the Brundtland Report, *Our Common Future*, 1987 (Brundtland 1987) introduced one of the major challenges of our time, the *sustainable development that meet the needs of the present without comprising the ability of future generations to meet their own needs.* Following the Brundtland Report, the *Rio Declaration on Environment and Development, United Nations Agenda 21 on Sustainable Growth*, 1992 (UN 1992), stated that human beings are at the centre of concerns for sustainable development. They are entitled to a healthy and productive life in harmony with nature (Principle No. 1). Not only were the limited resources a problem, but also the great problems of pollution thanks to earlier times unlimited inattention to the use of toxic substances in the resources. In the last decade of the nineteenth century, the term *Cleaner Technology* was launched, challenging both the need for facing the past use of toxic materials and the future use of materials that avoid toxic materials harming the environment. The other challenge of cleaner technologies was the preserving of resources by reduction, reuse, recycling and recovery (4R) of waste materials. The natural resources are limited, and the demand for resources is growing. The cost of exploiting the resources and the load on the environment during transport and utilisation of these resources has increased. Therefore, it is necessary to save resources and to use them efficiently with respect to sustainable development, climate change and the circular economy.

1.1.2 Sustainable development

Sustainable development is based on three pillars:

- Social sustainability
- Economic sustainability
- Environmental sustainability

Building and construction activities contribute to all pillars. Our social behaviour depends on our housing, which is the frame of our daily lives. The infrastructures provide communication and transport opportunities together with common utilities, which are necessary for the functioning of society. Besides providing food, building and construction activities are important contributions to the conditions of our survival. Depending on our capacity and our needs, and requirements for the buildings and infrastructures, we spent most of our income on our homes and communication. Currently, with the development of society, buildings and structures need transformation according to our needs. We talk about new and bigger houses and apartments following the need for more available housing space. New and broader roads and bridges for more traffic, and so on. All kind of building and construction activities affect the environment. The structures need space. During the construction process, this activity consumes energy and resources. It produces waste, noise, smoke, greenhouse gases and other impacts or inconveniences on the environment.

The *UN Sustainable Development Goals* (SDG) reflect the overall global need of saving resources. In 2000, the UN adopted the eight *Millennium Development Goals by 2015*. Goal no. 7: *Ensure environmental sustainability*, was aimed to integrate the principles of sustainable development into country policies and programs (UN Report 2015). The 2005 World Summit on Social Development identified sustainable development as the three above-mentioned pillars: economic, social and environmental development. The three pillars have served as a common ground for numerous sustainability standards and certification systems in recent years, which also have impacted on the building and construction sector, particularly in the materials and construction and demolition waste (CDW) area. The SDGs were revised at the summit in September 2015, where the world leaders adopted the 2030 Agenda for Sustainable

Development, which includes a set of 17 SDGs (UN General Assembly 2015). The SDG number 11: *Sustainable Cities, Communities,* and number 12: *Responsible Consumption and Production* address explicitly the need of saving resources for the construction industry.

1.1.3 Climate change

In December 2015, during the 21st Session of the Conference of the Parties in Paris, 195 countries adopted the first-ever universal, legally binding global climate deal. Besides the UN Sustainable Goals, we must also look at the climate change and reduce the emission of greenhouse gases (GHG) – *Green Change* – as stated in the Paris Framework Convention on Climate Change 2016, the *Climate Agreement (COP21)* (UN COP21 2015). Instead of consumption of energy and products made be non-renewable resources, consumption must be based on renewable resources. For instance, the energy consumption must change from non-renewable fossil sources, such as oil and natural gas, to renewable types of energy such as wind and solar energy. The agreement sets out a global action plan to put the world on track to avoid dangerous climate change by limiting global warming to well below 2°C. The agreement is due to enter into force in 2020.

The parties are committed to prepare national actions plans to ensure the control of greenhouse gas and achievement of these goals. The most abundant GHG in Earth's atmosphere are, in order,

- Water vapour (H_2O)
- Carbon dioxide (CO_2)
- Methane (CH_4)
- Nitrous oxide (N_2O)
- Ozone (O_3)
- Chlorofluorocarbons (CFCs)

The most relevant and important GHG to the building and construction sector, including the CDW sector is carbon dioxide, CO_2. Looking at the building and construction sector the CO_2 emission is a product of combustion and has a direct link with the energy consumption of machines and processes in the building production industry, as well as in the construction and demolition of buildings and structures. The goal of sustainable CDW management is the minimization of energy consumption and the resulting CO_2 emission, that is the production of cement, steel, aluminium and bricks. The overall goal is to reduce the emission of GHG to zero by 2050.

1.1.4 Circular economy

In all communities, it has always been common practice to retrieve valuable materials from the arising waste, such as metals and building materials.

After some decades in this century with an extensive 'use-and-throw-away' philosophy, it has been realized that we cannot continue this uninhibited use of natural resources and pollution of the world with waste. It is necessary to change our habits and to revise former common practices within the building and construction industry as well as other industries, households, and so on.

In the last decades, many *green* movements have arisen mostly based on political and idealistic issues, and unfortunately with a rather limited impact on practical life. However, within the last few years the World Bank, The Organisation for Economic Co-operation and Development (OECD) and the World Economic Forum (WEF) have emphasised that recycling of waste and introduction and implementation of environmental friendly technologies must be considered as one the greatest technological challenges of our time. To encourage the achievement of these objectives, the World Bank has faced the realities and clearly stated that improvement and protection of the environment is a question of money, which should be paid by the developed countries.

The *Circular Economy* is a generic term representing the overall vision on preservation the resources of the planet by recirculating resources and products. The consumption of resources and production of waste must change to lesser consumption of resources and substituting resources with recyclable waste. The key issues of the circular economy are economy, resources and CO_2, as presented in Figure 1.1.

Looking at the circular economy and resource efficiency related to the production and consumption of resources in the construction industry, a number of facts are listed in Box 1.1.

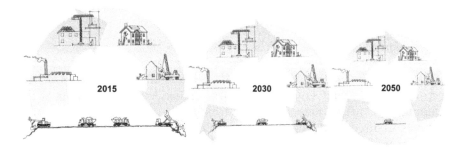

Figure 1.1 Principle sketch of the life cycle of buildings, which indicates the need for resources, production of materials and generating of construction and demolition waste (CDW). The 2015 scenario indicates the high demand for resources and CDW generation today. The scenario also indicates the general focus of opportunities for recycling waste materials. The 2030 scenario indicates increased focus and implementation of resource efficiency and circular economy according to UN 17 SDGs 2030. The 2050 scenario indicates the vision of full attention and implementation of resource efficiency, management of waste as resources and CO_2 reduction in accordance to the 2050 Goal of neutral CO_2 emission.

BOX 1.1   FACTS RELATED TO CIRCULAR ECONOMY AND RESOURCE EFFICIENCY IN THE CONSTRUCTION INDUSTRY

CONSUMPTION OF RESOURCES

- The construction and the use of buildings in the EU account for about half of all extracted materials and energy consumption.
- Construction materials by weight account for 44% of the resource used in the United States, 58% in the case of Japan and as much as 68% in Germany. The flow of construction materials is therefore responsible for a significant share of the global ecological footprint being 20% of the ecological footprint in the United States.
- More than 30%–50% of the total materials used in Europe are for housing. Around 65% of the total aggregates (sand, gravel and crushed rock) and approximately 20% of the total metals are used by the construction sector.
- The EU27 consumed between 1.200 and 1.800 million t of construction materials per annum for new buildings and refurbishment between 2003 and 2011. Concrete, aggregate materials (sand, gravel and crushed stone) and bricks make up to the 90% (by weight) of all materials used.

CDW GENERATION

- EU-28 generated 868 million t of CDW in 2017, 34.7% of the total waste generation. Population of 512 million, on an average 1.7 t/person.
- The United States generated 480 million t of CDW in 2012. Population of 310 million, on an average 1.5 t/person.
- Concrete, masonry rubble and stone stand up as the dominant components of 70%–90% CDW of which concrete made up approximately 50%.

RECYCLING

- EU target for recycling of CDW is 70% in 2020. However, the actual recycling in the member states vary from 0% to 98% of their CDW production.
- Recycled concrete aggregate is suitable for substitution of 10%–20% virgin aggregates for most concrete applications.

A typical life cycle of buildings and constructions consisting of following phases is shown in Figure 1.1:

- Supply of natural resources including excavating of natural resources, transport of resources to building sites or building materials production site
- Production of building material on industrial plants for the production of bricks, cement, doors, window, kitchens, etc.
- Construction on-site including infrastructure, foundation, building walls and roofs, closing the house with doors and windows, and completing the building with internal utilities
- Current maintenance during the use of the building including renovation, reconstruction and change of the building in accordance to the owner's need of the structure
- Demolition, deterioration or destruction and removal of materials

The basic problem with the model of circular economy and other similar models on circular thinking is the absence of the time history. The life cycle of buildings and structures, with estimated lifetimes of 50 years and more, consist of a mix of various processes of which the construction process is the most important. The processes and contracts during the lifetime of a building are based on time schedules from a starting date to a completion day – end of the contract. This means that we think, plan and act with a linear horizon. The circular mindset is an ideal and holistic strategy, which is the integration of the many operational linear processes, as shown in Figure 1.2.

1.1.5 Buildings and structures

Referring to the Danish philosopher *Søren Kierkegaard's* view of life, – *life is understood to be backward but must be lived forward* – the challenges of recycling and the sustainable approach of CDW management today must be based on the understanding of the use of resources in the past. The planning of construction, circular economy, recycling, and resource efficiency tomorrow must be based on the expected use of resources in the future. Figure 1.3 presents examples of recycling in ancient times.

In the early days, normal family buildings were built with cheap materials found on and around the locations. Nomads used mobile homes with materials of skin and sticks easy to mount, dismount and transport. Settled farmers and other people lived in permanent shelters and houses made of clay, wood, bamboo, stones and straw. Natural stones were found in the field and used as they were, or adjusted for foundations and walls. In areas without stones, dried mud bricks and peat bricks were used. Through the ages, iron, tile and mortar were introduced in the construction industry. Public buildings and wealthy people's houses were built with stones, brick,

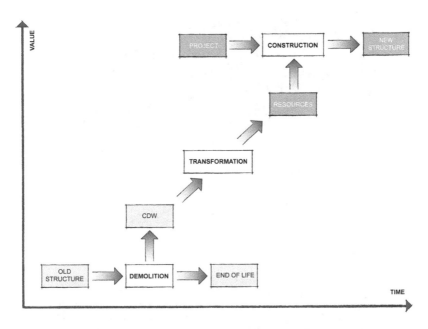

Figure 1.2 Principle of transformation of resources from one (old) life cycle to another (new), lifting the low value of CDW to a high value of resources.

wood and tiles. Block stones were broken in gravel pits by hand power and produced individually or in standard forms for construction of houses and bridges. Timber for buildings were cut from trees, bricks and tiles were made of burned clay and built in masonry structures and roofs. Stones were the most preferred and expensive materials. During the development of the new capital of Russia, Saint Petersburg, between 1703 and 1710, Tsar Peter the Great (1672–1725) ordered that new buildings in Moscow should be constructed by timber because stones were needed for buildings in Saint Petersburg[1]. From the end of the eighteenth century, the most important development in our time was the invention of concrete, especially steel reinforced concrete. However, it should be remembered that cement-like binders based on lime and volcanic ash (pozzolans) were already used by the Greeks and Romans in ancient times. After World War II, the reconstruction of war-damaged cities in Europe and the Far East together with the growth of population and demand for housing, new and rapid construction methods were developed. Prefabricated concrete elements were introduced in the construction industry. Environmental harmful compounds such as paint containing lead, asbestos materials in pipes, plates and insulation, and polychlorinated biphenyl (PCB) in joints and surfaces were common used.

---

[1] Experience from the rehabilitation of the Danish Embassy in Moscow. The embassy building looks like a stone house but it is built with timber.

Figure 1.3 Reused material in ancient structures. Top: Sidon Sea Castle, Lebanon, built by the crusaders in the thirteenth century with reused ancient marble columns as wall anchors. Bottom: Foundation in structures by Hadrian (AD 117–138), built with a mix of bricks and stones (Forum Romanum, Rome).

Figure 1.4 and Table 1.1 present the typical types of buildings and the commonly used building materials through the ages. From the 1950s to the end of the 1980s, asbestos and other environmentally harmful substances were common used. Therefore, the demolition of the buildings from this period requires special attention. From our time (2015–2020), we expect to build with clean and not environmentally harmful substances. However, the use of specific substances such as nano-materials need special attention.

Talking about sustainability and resources, we look at the renewable and the non-renewable resources. Stones, brick and tiles of required quality for

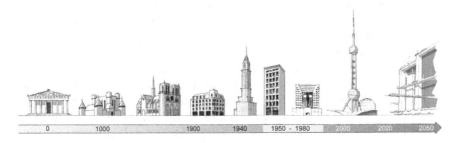

Figure 1.4 Development of building traditions through the ages. Until just after World War II, we considered the building materials as clean and not problematic. The period from the 1950s to the end of 1980s is highlighted because of the common use of asbestos and other environmentally harmful substances.

building materials are not renewable. On the other hand, stones and bricks might be imperishable and recyclable several times, presupposing that they are of good quality and handled carefully. They represent a certain value, including the cost of the production and the transport from the work to the building site. In principle, wood should also be considered as a renewable resource. However, we cannot replant new trees and match the consumption

Table 1.1 Typical construction materials through the ages

Time period, year	Characteristics of dominating construction materials
0–1900	**Clean and recyclable materials** Natural stones, bricks, iron, wood, glass, straw, mud, lime mortar and pozzolans cement mortar
1900–1950	**Recyclable materials** Natural stones, bricks, iron, wood, glass, plain and reinforced concrete, asphalt, lime mortar and insulation materials
1950–2000	**Recyclable materials and hazardous compounds** Plain and reinforced concrete, bricks, iron, aluminium, wood, glass, asphalt, lime and cement mortar, gypsum, mineral wool insulation, chipboards, laminated word, composites Environmental hazardous compounds e.g. asbestos, heavy metals, polyvinylchloride (PVC) and polychlorinated biphenyl (PCB), commonly used in the period from 1950–1980
2000–2020	**Clean recyclable materials** Plain and reinforced concrete, bricks, iron, aluminium, wood, glass, asphalt, cement mortar, gypsum, mineral wool insulation, electronic materials, chipboards, laminated word, composites and integrated construction elements
2020–2050	**Clean recyclable materials** Plain and reinforced concrete, bricks, iron, aluminium, wood, glass, asphalt, cement mortar, gypsum, mineral wool insulation, electronic materials, chipboards, laminated word, composites, integrated construction elements, 3D printed concrete, recycled materials, nanomaterials and other new materials (?)

of wood for construction buildings today, and it is very challenging for a society to maintain a rate of regeneration which reaches the rate of harvesting (timber felling).

According to the normal perception of a building owner, a building is a product, which starts with his need for a new home, office or another structure. The start of the building is based on functional needs expressed in a specific design and a financial investment. The construction of the building must fulfil the actual design and the intended use of the building. All buildings have a lifetime. Some buildings, for example private and public buildings, are used for many years according to their original design. Other buildings, for example production buildings and structures, are only used for a shorter period. In general, the lifetime of an ordinary building is expected to be 50 years. In Europe, a great part of the existing building mass is much more than 50 years old, while in the other part of the world the age of the building mass is very much lower. In the fast developing cities in the Middle East and Far East we are talking about life spans of 20 years and shorter. During the design of the structure, the inputs of resources are calculated in details. The scope of design does not comprise the future aspects and the possible transformation of the building during its lifetime. Therefore, the input as well as output of resources in the stages following the construction are not planned or calculated in details. Most building owners do not take care about the final output of resources needed by the end of a building's life.

The natural resources are limited, and the demand for resources is growing. The cost of exploiting the resources, and the load on the environment during transport and utilisation of the resources increases. Therefore, it is necessary to save resources and to use the resources efficiently with respect to sustainable development. The long – nearly infinite – lifetime and capability of construction materials for recycling makes a logical case for a closer look at the opportunities for a serious exploitation of recycled materials in new structures. This lead us from the linear lifetime thinking to a more circular lifetime thinking using CDW as a resource, thus substituting and saving natural resources.

1.2 AIM OF THIS BOOK

The aim of this book is to collect experiences and share information on sustainable development and circular economy in the construction industry with a special focus on the transformation of buildings, structures and resources. The transformation processes contain a chain of various processes related to the life cycle of buildings and structures presented in Figure 1.2. The overall challenges are the exploitation of opportunities for keeping the resources in the circuit at the highest quality level in the framework of best practice within the building and construction sector. To meet this challenge it is important to discuss and understand the following key issues of the transformation processes:

- Waste-to-resource dilemma
- Bridging the gap between the waste sector and the building and construction sector
- Comprehensive and integrated approach to CDW management and resource management
- Consolidation of state-of-the art and best practice of demolition and recycling

The circular economy focuses on keeping the resources in the circuit and saving resources, where the resource efficiency and the substitution of primary resources with secondary resources from building waste materials is a priority issue.

1.2.1 The waste-to-resource dilemma

Referring to the EU Waste Framework Directive (EU 2008), the EU defines waste as *'an object the holder discards, intends to discard or is required to discard. Waste is any substance which is discarded after primary use, or it is worthless, defective and of no use'*.

The UN definition is in line with the EU: *'Wastes are materials that are not prime products (that is products produced for the market) for which the initial user has no further use in terms of his/her own purposes of production, transformation or consumption, and of which he/she wants to dispose. Wastes may be generated during the extraction of raw materials, the processing of raw materials into intermediate and final products, the consumption of final products, and other human activities. Residuals recycled or reused at the place of generation are excluded'* (UN Glossary).

The dilemma is that in the name of sustainable development and circular economy the UN, EU Commission, US Environmental Protection Agency (EPA), and nations all over the world consider waste as a resource and try to impose this idea to common people. However, wastes are unwanted materials and resources are wanted materials. Waste is not a resource unless a certain transformation process has taken place depending on the type of waste and resources. Food and household waste can be treated biologically. Plastic, electronics, cans, cars, ships are processed to convert the waste to resources. The life cycles of most types of waste are relatively short – from days to weeks and a small number of years. Compared to other resources and products mentioned, buildings and constructions have a relative long life, and most of the materials are valuable and recyclable. From ancient cultures until our time, the world over, we see much evidence of reusing materials repeatedly for the same purpose. This is a substantial difference comparing to all other kinds of transitory products. The main components of CDW are concrete, bricks, steel, wood and glass, which can be recycled as follows:

- *Concrete*: Crushed and recycled as aggregate in new concrete or recycled as unbound materials in road construction
- *Bricks*: Cleaned and reused like new bricks or crushed and used as unbound materials in road construction
- *Steel*: Reused as construction steel profiles or scrapped and melted for the production of new construction steel
- *Wood*: Reused as timber or turned into woodchips for plates or gardening
- *Glass*: Reused or recycled for new glass production or production of insulation materials

Common to all types of waste is that they have to undergo special processes and procedures in order to be transformed into needed resources. The barriers for transformation from waste to resources are quality with respect to technical specification, risk of pollution and market value. Therefore, it is very important that we are familiar with the streams and all transformation processes from demolition and waste production to recycling and sale or disposal.

In order to facilitate the transformation of waste to resources we talk about an end-of-waste (EOW) criteria with documentation of the material quality and purity. The introduction and implementation of EOW criteria will contribute to the buyer's confidence in the material.

The waste-to-resource dilemma has an influence on the general debate about CDW management because the 'waste' word has a negative connotation compared to 'resources' or materials. In the United States, the EPA uses the words' construction and demolition materials, (CDM).

1.2.2 Bridging the gap

In the EU and the member states, the environmental legislation form the requirements for the management of CDW. The key issues of the legislation is the reduction of the impact of CDW handling into the environment, especially pollution by toxic substances, such as polychlorinated biphenyl (PCB), asbestos, and heavy metals. Local authorities establish the specific rules of sorting CDW into fractions, handling and reporting of contaminated materials, and so on. The legislation and rules for building and construction activities together are under the authorities of housing and construction, where the issues of recycling waste materials and substituting natural resources with secondary materials are not in focus. Therefore, the lack of incentives and commitment of the construction industry, as well as the gap between the authorities related to bringing recycled materials on the market are important barrier to recycling CDW.

The waste industry is linked to the environmental protection legislation and the construction industry is linked to the housing and construction legislation. Therefore, the typical stakeholders are split as shown in Table 1.2 and Figure 1.5.

The actors are the owners, the designers and the contractors. Traditionally, the building owners and their consulting architects and engineers do not take care of the demolition and waste management processes. Tendering of demolition contracts in most European countries is based on very basic information about the buildings to be demolished. The building owners and their consultants consider the demolition contractors as specialists who are responsible for all services of demolition and waste management, including recycling and disposal of CDW. Usually, the tender documents comprise descriptions of risks related to hazardous substances and the requirements for cleansing structures before demolition.

Due to the recent awareness of the requirements for sustainable development and circular economy together with the risks of pollution, many building owners have acknowledged their responsibility with respect to demolition and CDW management. In the case of sustainability certification according to LEED[2], DGNB[3] or BREEAM[4], the building owners and their consultants must assess the aspects of future demolition, dismantling and CDW management. Furthermore, it is expected in the future that the building owners and their consultants will assess the opportunities of recycling to a much higher degree than at present in cases of demolition, as well as the opportunities of using recycled materials in construction of new buildings. The acknowledgement of the economic and environmental benefits of recycling CDW materials, and the acceptance of recycled materials will bridge the former gap between the waste industry and the construction industry, as illustrated in Figure 1.5.

1.2.3 Comprehensive and integrated approach

The circular economy and the recycling of CDW require that we keep the resources in circuit from one building life cycle to another building life cycle. In case of urban renewal and development, for example old industrial sites, obsolete dwellings or harbour areas, which are often seen in western industrial countries, it is necessary to integrate the demolition and recycling activities with the development of the built environment. The integration must take place in long-term comprehensive planning and concerted actions.

The reuse of buildings and building materials is an important part of the construction industry and urban development. As mentioned earlier, the challenges of reuse of buildings and materials are dominated by a very high potential for building waste and a growing demand for reusable materials

---

[2] LEED: Leadership in Energy and Environmental Design, US Green Building Council.
[3] DGNB: Die Deutsche Gesellschaft für Nachhaltiges Bauen – DGNB, German Sustainable Building Council.
[4] *BREEAM*: British Research Establishment Environmental Assessment Method.

Table 1.2 List of stakeholders related to CDW management

Group of stakeholders	Typical stakeholders
Public authorities	*Regional, federal and governmental authorities* responsible for the legal framework and regulations, for instance national environmental protection agencies (EPAs)
	Local and municipal environmental authorities responsible for waste reporting, handling, permits and supervision
	Local and municipal housing and construction authorities responsible for building and demolition permits and supervision
Owners	*Owners of structures and buildings* and their consultants; architects, designers and engineers. The owners are responsible for the overall construction and demolition projects and the waste generation
Waste industry	*Waste management contractors* specialize in collection sorting and handling all kinds of waste, e.g. municipal household waste, industrial waste, hazardous waste, private building waste, etc. The waste management industry operates according to contracts with public and private organizations, building owners and industry
	Recycling companies specialize in preparing waste for recycling and selling recycled materials
Construction industry	*The raw material producers* excavate sand, gravel and stones from nature and provide sorted and sieved basic materials to other sectors. The raw materials industry provides great amounts of waste consisting of residual low-grade materials to be handled on-site by the producers. Recycled industrial mineral waste, e.g. fly ash and slags, and CDW materials can substitute some sand and gravel fractions of the natural materials
	The building materials industry produces all needed materials for the construction and building processes. The various processes generate many types of waste, which are considered as industrial waste. In most cases, the individual producers have developed methods and procedures of recycling their own waste. Some of the producers have developed recycling of their own products after demolition of buildings and structures. This is the case for some mineral wool producers, e.g. the Danish company Rockwool, gypsum plate producers, asphalt producers and glass producers
	The construction industry, infrastructures, is familiar with recycling concrete and asphalt in roads and airports. The use of asphalt or concrete depends on local accessibility of resources, traditions, traffic load and economy. In the United States, the Federal Highway Association has implemented recycling of concrete using crushed concrete as aggregates in new concrete road pavement since the 1970s (FHWA 2006). In Europe, the use of recycled aggregates has mostly taken place as unbound sub-base materials. In general, most of the resources from repair, renewal and demolition of bridges, roads, airfields, tunnels and other kinds of infrastructure are recycled and fulfil the objectives of sustainable development and circular economy

(Continued)

Table 1.2 (Continued) List of stakeholders related to CDW management

Group of stakeholders	Typical stakeholders
	The construction industry, housing, is the main contributor to the CDW from, repair, renewal and demolition of buildings. The waste production during the construction of buildings depends on the planning and management of the building processes
Demolition industry	*The demolition contractors* are specialized contractors, usually organized under the construction industry and working in construction as subcontractors. Some demolition contractors recycle and sell recycled materials while other demolition contractors subcontract recycling and market recycled materials
	Recycling companies specialize in recycling CDW, which comprise receiving and sorting CDW, and production of crushed masonry and concrete as secondary raw materials

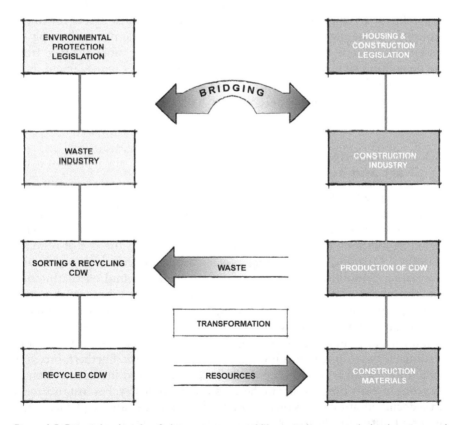

Figure 1.5 Principle sketch of the two sectors: Waste industry, and the housing and construction industry, along with the need for bridging them in order to promote transformation of waste to resources.

in order to substitute for natural resources. The opportunities for reuse of buildings and materials are based on overall economics and logistics management of resources of urban development projects. However, most buildings and structures contain substances which are potentially dangerous for the environment and human health. The risk of pollution caused by materials hazardous to the environment is a key barrier to the recycling of buildings and materials. Some buildings, as mentioned in Figure 1.4 and Table 1.2, have been constructed with materials containing substances considered harmful today, such as asbestos, PCB, heavy metals, certain paints, and so on, which can constitute a problem both during the 'normal' use and the renewal operations of the buildings.

Looking at the principle of circular economy and green changes, all stakeholders of demolition and construction (see Table 1.3) must make an early assessment in their projects to discus the opportunities for recycling building materials. All stakeholders should assess all opportunities to improve recycling and the use of recycled materials in new structures with a minimum of transportation, moving the materials from stockpile to stockpile. To implement successful matching, we must assess the linear processes of the demolition processes compared to the planned supply of resource for a new construction using the shortest possible distance from the demolition site. This means that two sets of project management must be involved: the building owner of the new building and all his consultants with the contractors on one side, and the building owner of the building to be demolished with his consultants and contractors on the other side.

1.2.4 Consolidation of state-of-the-art and best practice

The knowledge and sharing of experiences on demolition and recycling is an important precondition for implementation of demolition and optimal recycling. Much scientific literature exists on recycling of materials, including methods for recycling specific materials and the use of recycled materials. Demolition techniques are described in technical papers and fascinating demolition projects are shown in videos found on the Internet via YouTube. A lot of articles and presentations describe demolition of buildings and constructions, and management of waste for construction and demolition activities. However, very little international literature exists on linking demolition, recycling and marketing of the resources with respect to sustainable development in a broader holistic perspective. Furthermore, very little information is found on practical experiences and implementation of recycling in daily business life in the construction industry. For instance, very little documentation exists on the practical exploitation and implementation of opportunities for the use of crushed concrete as aggregates in new concrete.

The objective of this book is to describe and explain the basic principle for the management of output materials from construction and demolition

Table 1.3 Overview of the major fractions of CDW, EU-27 estimates, their distribution and potential of recycling, reuse and recovering. Rough estimates based on Bio et al. (2011) and other sources

CDW fraction	Type of material	Distribution min-max by weight %	Potential of recycling, reuse, recovering
Concrete, plain concrete or reinforced concrete	Inert, ceramics	50–90	Recycling of crushed concrete, bound and unbound applications
Masonry, bricks and tiles	Inert, ceramics	35–50	Reuse of bricks recycling of crushed materials
Stones	Inert, ceramics	1–2	Reuse of granite blocks, recycling of crushed aggregates
Asphalt	Mixed inert and combustible	1–5	Recovering and recycling in new Asphalt layers
Wood, timber, planks, doors, windows	Bio – degradable, combustible	1–10	Reuse of doors, windows, timber and planks, production of chipboards, energy recovery
Scrap iron, steel, metals	Inert, metal	1–5	Recovering
Gypsum		0,2–0,4	Recovering
Glass	Inert	0,1–1.0	Recovering
Insulation materials, mineral wool, glass wool	Inert	1–10	Recovering
Paper, cardboard, plastics	Combustible		Incineration, energy
Ceramics and porcelains	Inert		Disposal, recycling possible
Hazardous substances			Disposal or special treatment

activities and the use of these materials again for construction purposes. The aim includes:

- Collect and disseminate general information and experiences on CDW management
- Preparation of an overview of the life cycle and chain of value of buildings and structure planning to end-of-life, characterized by Figure 1.2 and the linkage mechanism
- Presentation of a comprehensive and easy-to-read book with simple ideas and guidelines about the most important processes concerning the lifetime of buildings and constructions, adopting the perspective of sustainable development and circular economy

The book shall inspire all persons involved in urban development, building and construction projects to think in terms of simple solutions about saving resources and money.

Demolition and CDW management is a global issue. Therefore, it is also the intention that the book should reflect the actual activities around the world, not only in EU member states, but also in all other regions and the most industrially developed countries. Natural disasters and conflicts all over the world generate huge amounts of solid waste, especially debris, and cause great challenges to the exposed societies. The basic principles of managing disaster waste are modelled on the principles of CDW management in industrialized countries. Therefore, integrated disaster waste management is an important topic in the book.

1.3 THE BACKGROUND AND LIMITATIONS OF THE BOOK

The book is based on the author's personal experiences since the early 1980s with demolition contracts and consulting engineering services around the world. The text of the book summarizes more than 100 technical articles and presentations on demolition, management of CDW and disaster waste management. It should be noted that much recycling experiences come from Danish research and development from 1986 to 1995, when the Danish Ministry of Environmental Protection implemented a number of action plans focusing on the promotion of recycling CDW. The achievements were remarkable. During this decade, the percentage of recycling CDW increased from 12% to approximately 90% (Lauritzen and Hansen 1997). However, today CDW management in Denmark does not differ substantially from CDW management in the EU and other industrialized countries. International experiences and activities from the following organisations, among others, also contribute to this book:

- The work within the RILEM[5] Technical Committees (TC): TC 37-DRC on Demolition and Reuse of Concrete, 1981–1988, TC 121-DRG on Guidelines for Demolition and reuse of Concrete and masonry, 1989–1993, TC 198-URM on Use of Recycled Materials, 1994–2004, TC-217 on Progress of Recycling in the built Environment, 2004–2015
- EU Framework Programmes for Research and Technical Development
- EU Commission actions on *A Resource-Efficient Europe*, focusing on CDW, raw materials and circular economy
- EU Construction & Demolition Waste Management Protocol

---

[5] RILEM: Réunion Internationale des Laboratoires d'Essais et de Recherches sur les Matériaux et les Constructions/International Union on Testing and Research Laboratories for Materials and Structures.

- American Concrete Institute ACI Committee No 555, Concrete with Recycled Materials
- European Demolition Association (EDA)
- The World Bank and United Nation organization programmes and guidelines on disaster waste management after disasters and conflicts in various countries, including Lebanon, Bosnia, Kosovo, Haiti, Nepal and others

Besides the research and development activities on recycling CDW within the RILEM organisation, the European Commission has initiated research and development programmes on basis of the EU waste policy stated in the Framework Directive on Waste Management (EU 2008). The EU has identified CDW as a priority waste stream. A number of Framework Programs for research and development on recycling CDW have been launched since the 1980s, including specific key actions and subprograms, for instance *Recycling Waste Research and Development* 1990–1992 (REWARD) and *Science and Technology for Environmental Protection* 1989–1992 (STEP). Of special importance with respect to the background of this book, the project *Integrated Decontamination and Rehabilitation of Buildings, Structures and Materials in Urban Renewal* 2003–2006 (IRMA) should be mentionned (IRMA 2006). For the moment, the Framework 8th programme is running under the name HORIZON 2020, which also funds projects of importance to recycling of CDW. The EU Commission launched a study from 2015 to 2016 on *Resource Efficient Use of Mixed Waste* focusing on CDW. The study comprised: (1) analysing the current CDW management situation in the member states (EU28), (2) performing six case studies of successful sustainable CDW management projects, (3) identifying good practice and formulation of a set of recommendations of CDW management, and (4) assessing the plausibility of official CDW statistics. Task 1 of the study resulted in a comprehensive study of all EU 28 member states, including the state of research and development of the individual member states (EU Study 2016).

Technologies for identifying and sorting hazard substances from CDW are being developed and put on the market. However, the technologies and methods are rather expensive. Technologies for recycling the major CDW fractions such as concrete, masonry and asphalt are developed and implemented. However, very little documentation on the technical and economic experiences or results is found. The major challenge is preparing procedures, including EOW criteria, which are necessary to bring the materials on the market as competitive alternatives to primary materials.

This book focuses on recycling traditional materials from the building and construction sector. It does not comprise recycling of waste derived from types of waste other than CDW, such as industrial waste and municipal waste containing paper, plastic, carpets, furniture and other kinds of waste prepared for the production of building materials. On the other hand, the book does not mention recycling of CDW for other purposes than building materials, nor for example, recycling wood and timber for furniture.

The pollution from building waste materials is a serious issue with respect to recycling and the acceptance of reused materials in new buildings and structures. However, here the pollution problems and the cleansing of materials are only presented in a general way. The book will not present the details on this issue. Please refer to the related literature.

1.4 DEFINITION OF TERMS USED

Communication is an overall challenge. CDW management comprises a lot of terms and definitions, which are not clear and uniform. In many cases, the terms cause confusion and misunderstandings. The EU Commission has established definitions presented in various directives and resolutions. The EU member states have their own definitions and interpretations of the EU definitions. USEPA has other terms. Therefore, it is necessary to present the most dominating terms and definition used in this book. Please refer to the comprehensive list of terms and definitions in the Glossary (Annex I).

1.4.1 Construction and demolition waste

Depending on their design, construction, use and maintenance all buildings are subjected to some kind of transformation during their lifetime, causing an input and output of resources. The output of resources during the construction, repair, reconstruction, maintenance of buildings and structures is called *construction waste*; the output of removed used material from repair, reconstruction and demolition is *demolition waste*. The two types of waste together are called *construction and demolition waste (CDW)*, falling under the construction and demolition waste categories referred to in the list of waste – EC decision of 3 May 2000 (EU 2000) adopted by the member countries pursuant to the EU Waste Directive (EU 2008). CDW is one of the heaviest and most voluminous waste streams generated in the EU. It accounts for approximately 25%–30% of waste generated in the EU.

The USEPA does not talk about waste but *construction and demolition materials (CDM)*. CDM consist of the debris generating during the construction, renovation, and demolition of buildings, roads, and bridges. CDM does often contain bulky, heavy materials, such as concrete, wood, metals, glass, and salvaged building components.

In many countries, industrial as well as developing, CDW is considered as harmless, inert waste, which does not give rise to problems. However, CDW consists of huge masses of materials, which often are deposited without any consideration, causing many problems and inviting the illegal deposit of other kinds of waste and garbage. Whether CDW waste originates from clearing after natural disasters or from human controlled activities, the utilisation of the waste by recycling can provide opportunities for saving energy, time, resources, and money. Furthermore, recycling and controlled management of CDW will save use of land and create better opportunities for handling other kinds of waste.

Through the processing, fabrication and transportation from the place of origin to the place of use the inert materials obtain value. The materials can be used repeatedly in many generations. Organic materials on the other hand have a certain lifetime and must be renewed or changed after a period. Whether we look at ancient building materials or modern sophisticated materials, the materials have a lifetime or a life cycle, depending on the materials and the use of the structure. However, CDW is special in that the public does not consider CDW as waste in the negative sense like rubbish. This is because most of it consists of concrete, bricks, steel and wood; materials which are not normally regarded as being hazardous. See the discussion on the waste-to-resource dilemma in Section 1.2 and Chapter 2.

CDW has been identified as a priority waste stream by the EU. There is a high potential for recycling and reuse of CDW, since some of its components have a high resource value. In particular, there is a reuse market for aggregates derived from CDW waste in roads, drainage and other construction projects. Technologies for the separation and recovery of construction and demolition waste are well established, readily accessible and in general inexpensive. In some EU member states, such as Denmark and The Netherlands, the percentage of recycling reaches figures close to and above 90%.

However, in order to avoid too much confusion reading this book, the overall term of CDW should be considered as waste, resources or materials arising from activities such as the construction of buildings and civil infrastructure, total or partial demolition of buildings and civil infrastructure, as well as road planning and maintenance. Different definitions are applied throughout the EU, which makes cross-country comparisons cumbersome. In some countries, even the materials from industrial production of building materials and land levelling are regarded as CDW. In this book, the fractions of CDW are defined in Table 1.3. Beside the fractions, Table 1.3 presents a brief description of the types of waste, their percentages in CDW according to European building tradition and opportunities with respect to reuse, recovery and recycling. It should be noted that soil and residue from excavating or production of raw materials are not included in CDW. Waste from production of building materials should be considered as industrial waste, and be excluded from CDW. However, in some cases the line between industrial waste and CDW is vague.

In Europe, waste generation of 10% of the total consumption of resources is expected (Ellen MacArthur Foundation 2015).

1.4.2 Disaster waste

Management of waste from disasters, in particular waste from destroyed buildings and structures is addressed in this book, because the principles of management and recycling are very much like the principles for management and recycling CDW. Disaster waste is defined as all kinds of waste from buildings and structures destroyed by natural disasters and conflicts. Large quantities of building waste are generated within a very short time period comprising concrete debris (rubble) masonry, wood, steel and other materials. Other kinds of solid and organic waste should also be taken into consideration. However, these are outside the framework of this book.

1.4.3 End-of-waste

End-of-waste (EOW) status: When materials characterized as waste have undergone a recovery, including recycling and operation, and comply with specific EOW *criteria* to be developed according to the EU Waste Directive:

- The substance or object is commonly used for specific purposes
- A market or demand exists for such a substance or object
- The substance or object fulfils the technical requirements for the specific purposes and meets the existing legislation and standards applicable to products
- The use of the substance or object will not lead to overall adverse environmental or human health impacts

The criteria shall include limit values for pollutants where necessary and shall take into account any possible adverse environmental effects of the substance or object. The EU Joint Research Centre (JRC) rapport End-of-Waste Criteria, EUR 23990 EN-2009 provide guidelines for EOW criteria for CDW.

1.4.4 Reduction, reuse, recycling and recovery (4Rs)

The terms *reduction, reuse, recycling* and *recovery* are often problematic to understand and separate. In Annex I, the terms are defined in accordance to the EU Waste Directive (EU 2008). In this book, the term *recycling* is used as a general term for *recovery, reuse* and *recycling* and all related transformation processes from demolition to the use of materials including collection, sorting, treating and preparing for use.

1.4.5 Recycling hierarchy

In the model of circular economy, the recycling aims at the highest potential of material quality, characterized by a cascade of solutions presented as the *waste hierarchy* in the EU Waste Directive (EU 2008). The *waste hierarchy* (see Figure 1.6) provides a priority order in waste prevention, management legislation and policy (Table 1.4).

1.4.6 Resources, raw materials

The demand of resources and products for buildings and construction comprises an infinite amount and types of materials, of which some materials are very suitable for recycling, while the potential for recycling of other materials depends on various parameters. The most typical potential for recycling is raw materials such as gravel and stones excavated from the sea and land, and metal produced from steelworks. Raw materials are divided into two categories:

- *Primary raw materials*: Virgin or natural materials/resources
- *Secondary raw materials*: Recycled, recovered or reused materials from construction and demolition processes with technical properties enabling to substitute primary raw materials

1.4.7 Transformation

In this book the term transformation is a key word. We talk about transformation processes with respect to building, structures and materials

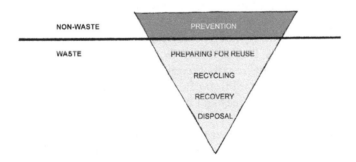

Figure 1.6 Hierarchy of waste according to the EU Waste Directive (EU 2008).

Table 1.4 Hierarchy of CDW materials

Level	Quality
1.	Reuse of buildings
2.	Reuse of building components and materials
3.	Recycling of metals, wood, concrete, asphalt, glass
4.	Backfill of crushed materials
5.	Disposal, special treatment

changes from one life cycle to another, and from waste to resources. The critical stages during the transformation processes are end-of-use, end-of-life and end-of-waste (EOW). See specific criteria for EOW in Section 1.4.3.

1.5 HOW TO READ/USE THE BOOK

1.5.1 Why read this book?

The book has a practical aim. In general, it addresses all persons involved in the processes of transformation of structures, management of resources and CDW: building owners, planners, consulting engineers, architects, demolition contractors, general contractors, civil servants, authorities, universities and education institutes, networks, and so on. The book focuses on transformation of waste to resources and resource efficiency in the construction industry, especially the management of resources and supply of resources from CDW substituting the consumption of natural resources.

CDW management and recycling operations take place in the construction industry during an endless number of various scenarios. The book deals with the following three representative scenarios:

- *Single-building transformation* of old buildings to reused buildings. Partial demolition and reconstruction of existing buildings
- *Single-building transformation* of old buildings to new buildings. Demolition of buildings after end-of-life, recycling of materials and construction of new buildings
- *Multiple-building transformation* involving transformation of several buildings from old city to new city. Urban development and renewal, including repair, reconstruction of existing buildings and infrastructures, demolition of existing buildings and infrastructures, recycling and use of recycled materials on-site

Next after saving lives, clearance of solid waste is the most important action after natural disaster and conflicts. Materials from damaged buildings and infrastructures should be handled as resources. Therefore, it is important for international non-government organizations (NGOs) and other aid organisations to know and exploit the experiences on demolition and recycling in industrial countries.

Depending on the scenario, the building owners, architects, designers, planners and contractors of new buildings and demolition projects can learn about the following issues among others:

- Assessment of the potential of reuse materials from the existing building for use in the new buildings
- Assessment of the opportunities for recycling materials on-site

- Life cycle assessment (LCA)–based approach to CDW management
- Design of new structures for disassembly and recycling
- Minimizing cost of demolition and material management by identifying valuable materials and exploitation of opportunities for selling recyclable materials
- Use of recycled materials for reclamation of the site after demolition
- Optimal planning of the demolition process for giving the contractor the most advantageous conditions for minimizing his offer
- Matching generating CDW from demolition at one site and the need for resources for new structures at another site
- Economic and environmental models
- Planning of demolition projects in due time before the start of construction of new buildings
- Logistic planning and coordination of demolition work with the start of construction work
- Proper planning and timing of demolition project in order to save money and reducing risks
- Debris management after disasters

1.5.2 Presentation of the book

After this introduction, Chapter 2 presents the overall framework for sustainable use of materials in the construction and building sector. The chapter starts with the presentation of the life cycle of structures and the elements of the life cycle followed by a brief presentation of a life cycle assessment (LCA). A general description of CDW and its impact on society together with visions of the EU are described. The chapter presents and explains the key issue of this book: *transformation*. The description of transformation scenarios is highlighted with a number of practical examples of reuse of old buildings and urban development in a historical view. The chapter concludes with an overall presentation of one of the most important challenges of the transformation of waste to resources, namely the risk of contamination because of hazardous materials. The most important hazardous compounds are briefly presented.

Demolition, including planning, design, contracting and implementation of contracts is described in detail in Chapter 3. It presents the methods of selective demolition and the principles of environmental sanitation, cleansing of buildings and the separation of hazardous materials building materials. Great importance is attached to the management of CDW materials. The chapter concludes with a summary of technologies for concrete demolition and demolition methodology.

Chapter 4 presents the state-of-the-art for recycling concrete, masonry, asphalt and other materials. Opportunities for recycling and overcoming various barriers for recycling are highlighted. Bringing secondary resources, especially secondary raw materials to the market and user confidence in the materials are discussed. At the end of the chapter, examples of full-scale recycling projects, followed by a global overview of the recycling market are presented.

Chapter 5, on integrated CDW management, presents the holistic and optimal processes of demolition and recycling CDW, focusing on matching recycled resources from demolished buildings and structures with the need for resources for new buildings and structures. New concepts on single-recycling and multiple-recycling are introduced and explained. The city concept, developed through the EU-funded IRMA project, and practical demonstration projects are then briefly presented.

Disaster waste management (DWM), especially management of debris around the world is an important issue. Thus referring to Chapter 5, integrated DWM is implemented with respect to management of solid waste after natural disasters and conflicts, which is discussed in Chapter 6. The chapter focuses on integrated DWM, with respect to recycling and substituting secondary resources in the reconstruction processes. Based on experiences from a number of disaster and conflict scenarios in Bosnia, Kosovo, Lebanon, Haiti and Nepal, the chapter presents operational and economic models for debris recycling.

Finally, in Chapter 7, visions and opportunities together with proposed strategies for prevention and minimizing CDW complete the book. Innovative steps aimed at future designs for dismantling and recycling of buildings and building materials are presented.

The terms and synonyms glossary are found in Annex I and abbreviations are found in Annex II. References are listed at the end of each chapter, and Annex III presents a list of all references. The references are mentioned in the text in brackets with the name of the author/organisation and the publication year.

Chapter 2

Transformation of structures and materials

> All things in the World are a gift from the creator and must be considered as holy. In principle, things cannot disappear or be wasted.
>
> *Akio Morita, Founder of SONY*

2.1 INTRODUCTION

The context of recycling CDW and transformation of waste materials into resources is based on the life cycles of buildings, structures and materials. The perception of the context is anchored in the understanding of the linkage between the processes or elements of the life cycles. In this chapter, the life cycle thinking expressed in circular economy, resource effectiveness and CDW management will be discussed. The framework of CDW generation and management is split into three general transformation scenarios: single-building transformation scenarios (two scenarios) and a multi-building transformation scenario. Here, a number of examples of the three transformation scenarios are presented and reviewed.

The streams of materials in the life cycles of buildings and materials will be described together with the challenges and opportunities to control the streams. Decontamination of buildings and materials, which is a major challenge of transformation, will also be reviewed.

2.2 LIFE CYCLE OF STRUCTURES

The life cycle of buildings and structures presented in Figure 1.1 illustrates the circular streams of resources and waste. The figure corresponds with the stages of the life cycle defined by the European CEN Technical Committee 350 and standards EN 15978 and EN 15804. The standards divide the life cycle into four stages, as shown in Figure 2.1.

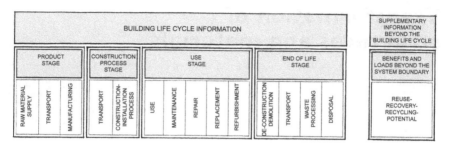

Figure 2.1 Building stages of the life cycle according to EN 15978-2011. (Credit: Danish Standards Foundation.)

- Production stage
- Construction process stage
- Use stage
- End-of-life stage

The building life-cycle stages presented in Figure 2.1 are generic and applicable for typical buildings or structures from cradle to grave. This type of life cycle might be useful for general LCA focusing on energy and greenhouse gas (GHG) emissions, environmental loads, and so on, where the long period of use stage provides the major impact. Looking at the material streams from a resource and CDW management point of view, the production stage, construction process stages and end-of-life stage are the most important stages. Compared to the input of resources and waste generation of these stages, the amount of material consumption and waste production during the use stages is marginal. The EN 15978 considers reuse, recovery, recycling potential as benefits and loads beyond the system boundary, which is not appropriate with respect to the concept of circular economy that keeps the resources in circulation. The idea of circular economy and transformation of CDW into demanded resources must be based on the linkage between various building cycles, where the materials transform from waste of one building cycle to resources in another building cycle, as presented in Figure 2.2.

In urban renewal and many other projects, the first stage of the building life cycle is demolition of the former structure on the ground. Most projects start with demolition creating space for new buildings and structures. The realistic life cycle should start with demolition/recycling and end with demolition/recycling.

In order to create the framework of a *material transformation system*, and define the basic parameters and indicators, a closer look at the individual processes and stages is needed.

When talking about the life cycle of a structure, we must remember that a complex structure like a house comprises a wide range of building parts

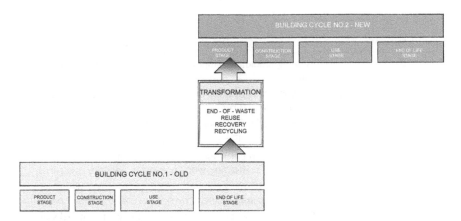

Figure 2.2 Linked building cycles enabling transformation of waste from building no. I (old) to resources for building no. 2 (new).

with various life cycles. Some parts have a short lifetime and need to be changed or renovated several times during the lifetime of the building. Other parts are more durable or do not need repair or changes. According to *Building a Circular Future* by Jensen and Sommer (2016), the following lifetimes for building parts are estimated:

- Foundations: 100+ years
- Structures, bearing parts of the construction: 50+ years
- Facades, not bearing parts: 30+ years
- Partitions and systems, internal walls, kitchen, toilets, etc.: 10+ years

2.2.1 Product stage

Minerals, aggregates, metal, wood and other resources are extracted from nature through the use of hand tools, mechanical equipment and blasting. Some materials are sorted in specific fractions and delivered to building material manufactures, such as brick works, cement and concrete factories and sawmills. Other materials are extracted for production of steel, glass, gypsum, doors, windows and many other types of building materials. A substantial part of sand and stones, excavated from land and sea goes directly to the construction sites.

The input of needed resources for excavation of the materials from nature during the product stage comprises various kinds of heavy equipment, crushing and sorting plants, human resources, energy and water. The input of resources for the manufacturing of steel, lumber, cement, doors, windows and other products requires resources depending on the manufacturing processes. Manufacturing of bricks, steel, glass and cement require high-energy processes.

Table 2.1 List of typical construction work and activities

Construction work	Typical activities and materials
Worksite	Fencing, shielding, temporary office and work facilities, parking sites
Clearance of worksite	Clearing of vegetation and trees, demolition and removal of former structures
Earth work	Excavating for basement, foundations and utilities
Foundation work	Piling, formwork, concrete work, insulation work
Structural façade and bearing structures	Formwork, in situ concrete work, setting up concrete elements, scaffolding, work platforms, security measures
Roof work	Bearing structures concrete, steel, wood, covering bricks, tarpaper, plates, insulation work scaffolding, insulation work
Interior works	Structural concrete and masonry walls, partition walls of wood, gypsum, lightweight concrete
Façade work	Windows, doors, façade covering
Installations an utilities	Heating, water, electricity, toilets, bathrooms, kitchens, ventilation, gutters, etc.
Finishing	Painting, gardening, outdoor work

Besides the consumption of materials for production of building materials, certain amounts of waste are produced. Discarded materials during extraction of raw materials are disposed of on the extraction or treatment site. Waste materials from the industrial production of cement, concrete, bricks, doors, windows, and so on, are recycled or reused by the producer, or handled by the waste management contractors as industrial waste. However, the streams of waste from the building materials industry have only a limited influence on the management of the major streams of waste materials from the construction and demolition processes.

2.2.2 Construction process stage

The construction work is tendered and contracted based on the design and planning processes.

The typical construction processes comprise the following activities, among others, as listed in Table 2.1.

The typical input of resources and output of waste is the following:

- The establishment of the worksite needs space, gate and fencing materials, office and workshop facilities, sand and gravel for roads and surfaces. Fences and shield materials, including concrete footings, are often reused materials from former works. Today, containers and reusable, mobile sheds are used. Only marginal amounts of wastes are expected from a modern worksite. No need for specific recycling activities

- The worksite clearance might result in considerable amounts of waste depending on the former use of the space. In case of demolition of buildings, CDW waste must be managed in accordance to the specific waste regulations
- The earthworks need earth moving machines, sand and stones, trucks for transportation and human resources. Often, there will be a great need for transport and disposal of excavated earth
- The foundation work depends on the geotechnical characteristics of the underground and the designed structures. The need resources are concrete piles and other concrete structures. The waste materials are mainly concrete and steel. In case of ready-mix concrete supply, surplus of concrete is expected to be tipped of on the site or another place
- The aboveground structure including the structural façades, bearing structures and the other construction work need all kinds of building materials

The construction waste materials are generated because of

- Over-ordering of materials to meet the risk of shortages of materials
- Discarded and faulty materials
- Inappropriate design
- Materials from shortening, adjustment, etc.
- Packaging materials

Normally, the individual contactors will handle their own waste materials, or a specialized waste management contractor is hired to handle the waste materials. In many EU countries, waste materials are collected at central waste collection points with marked containers for sorted waste. Some of the waste materials, for example mineral wool, gypsum, asphalt, steel, aluminium, plastic pipes, and so forth, are taken back by the producers and recycled. Packaging materials paper, plastic and wood may be incinerated for energy recovery.

Danish experiences have shown that normal construction work generates waste material on the order of 5%–15% of the total consumption of building materials. Referring to Table 2.3, according to the Ellen MacArthur Foundation (2015), 10%–15% of building materials are wasted during construction. It is very clear that this is an open window for improvement, such as for careful designs with respect to resource consumption and logistic management.

2.2.3 User stage

While the user stage is the dominating stage with respect to energy consumption the needed requirements of materials is limited. We are talking about materials consumed during running maintenance and repair work. Depending on the building, the lifetime of the building parts, and the users, the need for refurbishment and major structural changes of the buildings occurs one or more times during the expected lifetime of the building. The amount of CDW depends on the structural intervention and demolition work. Besides CDW, the waste management comprises construction waste similar to construction waste from new buildings.

2.2.4 End-of-life stage

The end-of-life stage is characterized by deconstruction/demolition and waste management, which are the major topics of this book described in detail in the following chapters, and comprises the following activities:

- Cutting all utilities, electricity, water, outlet, data connections, etc.
- Removal of abandoned furniture goods and installations
- Removal of hazardous materials contaminated with asbestos, PCBs, heavy metals and other polluting substances
- Selective demolition of the building starting with the removal of doors, windows, roof, light interior structures and partitions followed by demolition of the remaining structure
- Removal of foundations
- Clearing and rehabilitation of the site

The waste management comprises following activities:

- Handling hazardous waste, special treatment or depositing
- Sorting materials in fractions, such as concrete, masonry, stones, wood, steel, paper, plastic, etc., according to the specific regulation of the public authority
- Preparation of materials for reuse, recycling and recovery
- Incineration energy utilisation of combustible waste

The end-of-life stage and processes of reuse, recovery and recovery should be integrated in the material transformation management system. It should also be noticed that the end-of-life stage does not necessarily mean the structure will be deconstructed/demolished after end-of-use. Many buildings stands several years after end-of-use and decay before they are demolished as a part of a development project.

Table 2.2 presents the overview of resources, materials, waste and transformation of waste into demanded materials in accordance to the building life stages in Figures 2.1 and 2.2. Table 2.3 shows that the Ellen Mac-Arthur Foundation, in *Growth within a Circular Vision for a Competitive Europe* (2015), found that 54% of demolition materials in the EU are landfilled, while some countries only landfill 6%.

Table 2.2 Overview of resources, materials, waste and transformation potential of the individual life-cycle stages

Stage	Resources input	Materials output	Waste handling	Transformation
Product stage				
Raw material supply	Natural resources Hardware, energy	Sorted products	Discarded materials disposed on-site, no specific waste handling	None
Transport	Hardware, energy			
Manufacturing	Industrial resources, natural resources Energy	Building products	Industrial waste related to the products handled by the producer or waste management contractor	Recycled or reused by the producer or waste management contractor
Construction				
Transport	Hardware, energy			
Construction installation process: Foundations, façades, walls, roof, floors, interior, installations, etc.	Construction resources, materials, water, energy	New structures building ready for use Construction waste	Various waste fractions from superfluous and discarded materials, packaging, etc., handled by the contractors or waste management contractor	Recycled, recovered, incinerated or disposed
Use stage				
Use	Energy, water	Marginal	Marginal	None
Maintenance	Various resources	Marginal	Marginal	None
Repair	Various resources	Marginal	Marginal	None
Replacement	Various resources	Marginal	Marginal	None
Refurbishment, Depending on the structural intervention	Some resources	Some	Some	Some
End-of-life stage				
Deconstruction, demolition	Hardware, energy	Sorted demolition waste materials	CDW management	High 3R potential
Transport	Hardware, energy			
Waste processing	Hardware, energy	Sorted clean materials	Discarded and polluted materials, Landfill	
Disposal	Hardware, energy			

Table 2.3 Life-cycle figures related to construction and CDW management in the EU construction industry

Construction	Utilisation	Usage	End-of-life
10%–15% of the building materials wasted during construction	59% of European offices are not used even during working hours	20%–40% of energy in existing buildings can be profitably conserved	54% of demolition materials are landfilled, while some countries only landfill 6%
0%–0.5% productivity increase per year in most European countries 1990–2015, whereas 2% per year achieved in some countries	50% of residential dwellers report living in too much space	Passive building standards at or near profitability for most new build segments, but still only constitute a minority of buildings	Most materials are unsuitable for reuse as they contain toxic elements

Source: Ellen MacArthur Foundation (EMF). Delivering the circular economy – A toolkit for policy-makers, 2015.

Figure 2.3 presents an example of the variation of the user value of a building during the life cycle. Following the building owner's decision on starting the building project at time T1, the design and preparation of the project including tendering and contracting take some months. After completion, for example 12 months, the new construction is handed over to the user at time T3. Normally, the expected average lifetime of housing buildings is set to 50 years. The expected use time of industrial and office buildings are lower. During the use stage, the building will be managed and maintained with periodical upgrading and refurbishment, shown at time T3.1 and T3.2. The use stage and time of life of the building depends

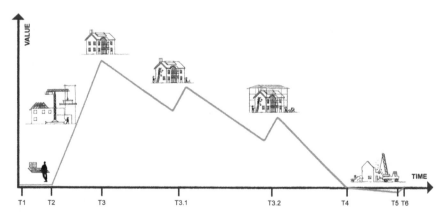

Figure 2.3 Example of the value of a building during a typical life cycle.

on the quality of the building and the actual demands for the building with respect to the original purpose, as well as other purposes. In the past decades in Europe, it is often seen that old industrial buildings are transformed to other kinds of industry, offices or dwelling use. However, the typical scenario is that the building, after a lifetime of 20–50 years, needs to be replaced with a newer and higher building. This is often the case in rapidly developing regions, typical of Middle East and Far East cities. The use stage ends at time T4, and the worn down building is left or sold/rented for other unspecified purposes. The maintenance is stopped and the building decays until a developer takes over the site for a new building project and the old building is demolished. By the end-of-use, most buildings are in bad condition, and might often be widely polluted with hazardous materials, such as asbestos, PCB, lead in paint, and so on. The value of the building is negative corresponding to the cost of sanitation and demolition.

Looking at the life cycle of a building, the most relevant stages with respect to materials, building waste, the time of planning and construction is typical 1–2 years, and the time of demolition typical 0.5 to 1 year; in total around maximum 2%–5% of the total lifetime. Looking at the cost of building construction in Europe today (2016), it is roughly estimated to be about 1000–2000 EUR per square metre, while demolition costs are roughly estimated to be 100–200 EUR – equal to 10% of the construction cost.

Regarding sustainable development, greenhouse gas emissions and the circular economy, the stakeholders in the building and construction sector focus on the user stage as the most important with respect to operation and maintenance cost, environmental loads, energy consumption and greenhouse gases. Looking at the materials, the construction and demolition waste, the potential of transforming waste into demanded resources, we must concentrate on the production and construction stages and the end-of-life stages.

Professor Charles Hendriks introduced *Chain Management* (Hendriks et al. 2000), which refers to the management of the entire life cycle of building materials, that is the complete chain of production, construction, demolition and reuse/recycling or disposal. Compared to the chain management and other types of life-cycle–based management approaches together with many new models of circular building economy, the integrated CDW management focuses on the actual transformation of structures and the transition of CDW into resources.

2.3 LIFE CYCLE ASSESSMENT

In a holistic approach, we consider all relevant perspectives of the sustainable development elements: economic, environmental and social issue. The LCA methodology primary deals with environmental impact, described in the EU guidelines (EU JRC 2010). However, as mentioned in the guidelines, the LCA can be combined with cost-benefit analyses.

The holistic approach for CDW management has a number of dimensions, for instance:

- Time history of buildings and structures
- Changes of culture, design and building style through the lifetime
- Multidisciplinary interests

Depending on their purpose and use, the lifetime of buildings and constructions vary. Some buildings and structures have a short lifetime of very few years, for example temporary structures for specific products. Other buildings and structures exist for hundreds of years. During the lifetime of the structure (see Figure 2.3), the structure will undergo maintenance and repair, eventually renovation and lifetime extension. By the end of the original purpose of the building, the building owner will face the question and decision on renovating or demolition and replacement with a new structure. The building owner's decision will be based on a set of decisive elements, related to the three key elements of sustainability, economy, environment and social impact. Referring to the three other cornerstones, the Corporate Social Responsible (CSR) building owner analyses the situation and the opportunities based on his/her intentions and financial goals.

LCA is an internationally standardized methodology in accordance to *ISO 14040 Environmental Management – Life Cycle Assessment – Principles and Framework*. LCA helps to quantify the environmental pressures related to goods and services (products), the environmental benefits, the trade-offs and areas for achieving improvements, taking into account the full life cycle of the product. Life Cycle Inventory (LCI) and Life Cycle Impact Assessment (LCIA) are consecutive parts of an LCA, where:

- LCI is the collection and analysis of environmental interventions data (for example emissions to air and water, waste generation and resource consumption), which are associated with a product from the extraction of the raw materials through production and use to final disposal, including recycling, reuse, and energy recovery
- LCIA is the estimation of indicators of the environmental pressures in terms of for example, climate change, summer smog, resource depletion, acidification, human health effects, etc., associated with the environmental interventions attributable to the life cycle of a product

The data used in LCA should be consistent and quality assured, and reflect actual industrial process chains. Methodologies should reflect a best consensus based on current practice.

From the conflicts around the world over the past years, we have learnt the terms of comprehensive approach – strategy for facilitating interaction between civil and military goals. Looking at the waste sector as the provider of recycled products, and the building and construction sector as the user of the recycled products, we should learn from the military's comprehensive approach to reach the unified goals on resource efficiency.

2.4 CONSTRUCTION AND DEMOLITION WASTE

2.4.1 Types of CDW

CDW is the major part of all waste production, in general 30%–40% of the total waste production in industrialized countries. CDW accounts for 34% of the urban waste generated within OECD countries. CDW is generated during the construction, renovation or demolition of buildings and structures, as mentioned above. Referring to Danish experiences (Lauritzen and Jacobsen 1991), the general figures of generated construction and demolition waste expected per square metre are:

- 23 kg, construction of new buildings
- 50 kg, renovation of buildings
- 1625 kg, total demolition of buildings

According the *Handbook of Recycled Concrete and Demolition Waste* (Pacheco-Torgal et al. 2013) estimated figures, kg CDW per square meter, are:

- 120.0 kg, construction of new buildings
- 338.7 kg, rehabilitation
- 1129.0 kg, total demolition
- 903.2 kg, partial demolition

The *Handbook* mentions information obtained from building permits in Thailand, with figures of 21.38 kg/m^2 waste for the construction of dwellings and 18.3 kg/m^2 waste for the construction of non-dwellings. The figures show great variations in estimation. This is probably because of the various methods of calculation and measuring used.

Data on CDW production are not collected routinely or consistently, so most published figures are estimates, which need to be interpreted with caution. Such estimates include 868 million tonnes of CDW generated across the EU-28 in 2017, 277 million tonnes in Japan, and 480 million tonnes in the United States (see Box 1.1). In EU member states, the reported CDW arising per capita vary in average from 1–2 tonnes/capita. The recycling rates of CDW range enormously among countries. The 2011 Bio report (Bio 2011) provides a 'best estimation' of the 2008–09 EU average in the range of 30%–60%, with EU countries reporting recycling and recovery rates as high as over 90% and as low as 10%. In the United States for 2009, approximately 40% of the CDW generated was reused, recycled or sent to treatment plants. On the long term, generation of CDW in 2020 around 520 million tonnes is expected.

The European Commission (EC) has identified CDW as a priority stream because of the large amounts generated and the high potential for reuse and recycling embodied in these materials. Indeed, proper management would lead to an effective and efficient use of natural resources, and the mitigation of the environmental impacts to the planet. For this reason, the EC Waste Framework Directive requires EU member states to take any necessary measures to achieve a minimum target of 70% (by weight) of CDW by 2020 for reuse, recycling and other material recovery, including backfilling operations using non-hazardous CDW to substitute for other materials.

2.4.2 CDW prevention

The EC Waste Directive gives definite rules for the management of wastes. The EC hierarchy of waste prevention is presented in Figure 1.6, and the hierarchy of CDW is listed in Table 1.4. The overall principle of CDW prevention is high quality and long life of new construction, minimization of the waste production during the construction processes and the highest quality of recycling CDW.

Some of the more important principles and rules are:

- The polluter pays the principle
- The polluter must report all waste to local authorities and authorities must assign ways and means of removal
- Handling of waste must be planned and goals and strategies for maximum possible recycling must be established

2.4.3 CDW recycling indicators

Usually CDW is measured in metric tons. It calls for attention when the recycled materials are brought on the market, because raw materials, such as stones and sand are measured in cubic metres. Recycling goals are expressed in percentage of recycled materials of the total generated CDW material regarding to the weight. The EC goals for recycling CDW are 70%, and many member states have much higher recycling rates. Considering the amount of recyclable materials in the existing building mass, a recycling goal of 95% is realistic. We know that more than 90% of buildings and structures are made of reversible materials, typically concrete and bricks. In principle, we should aim at 100% recycling, but a certain amount of hazardous CDW materials must be foreseen from the demolition of buildings, especially buildings constructed before the 1980s. It should be noted that the Ellen MacArthur Foundation finds that most materials are unsuitable for reuse, as they contain toxic elements (see Table 2.3).

The recycling percentage alone is not an applicable goal in itself. The goals of recycling should concentrate on the qualities of recycled materials with reference to the waste hierarchy and the potential substitution of natural resources or products. For instance, in Denmark it is a realistic goal to substitute natural gravel with at least 10% recycled crushed concrete. In addition, substitution of 10% of the total production of bricks with reused bricks is a realistic goal. A combination of the substitution percentage and the recycling percentage could form a relevant measure of the recycling goals.

2.5 TRANSFORMATION

The concept of circular economy keeps the resources in never ending loops at the highest level of value/quality as possible. In the building and construction sector, we talk about life-cycle spans of 20–100 years, on average 50 years. This means the input of resources in the beginning of the life cycle might have another character and value by the end of the life cycle, and that we are dealing with subsequent individual life cycles of materials, as well as buildings.

2.5.1 Transformation scenarios

Based on the coupled life cycle model shown in Figure 2.2, general models of the three transformation scenarios of old buildings into new buildings are discussed:

- Single-building transformation from old building to reused building
- Single-building transformation from old building to new building
- Multiple-building transformation involving transformation of several buildings from old city to new city

The general transformation models describe the idealistic scenarios, where the buildings are reused or demolished. The waste materials are transformed into resources, which are needed for a new building on the site, or another nearby. The typical scenario today is demolition of a building and waste management without specific plans for recycling. However, the various sustainable certifications of new buildings for example BREAM, DGNB and LEED give credit points to demolition and CDW management. The waste materials for recycling are transported and stocked on designated storage sites, or handed over to recycling companies and sold as secondary raw materials. The ideal scenario is when the reused building and its materials produced match the need for new buildings or materials. The following four preconditions are required:

1. The buildings and materials must be clean, which means they cannot contain any toxic substances/hazardous materials according to local environmental legislation and threshold values, according to specific EOW criteria

2. The materials must be sorted, prepared for recycling and substitution of new materials, typically natural primary materials that comply with specific technical standards
3. The cost of the material, including the total transformation cost, must correspond to the prices of natural materials delivered on-site
4. The material must be available in sufficient quantities at the right time

2.5.2 Reuse of buildings

The decision on reuse of a building or structure depends on the quality, history and suitability for new purpose. Many old historic buildings with cultural and/or architectonic qualities are preserved and subjected to restrictions with respect to changes. The problem is often to find a useful purpose for old preserved buildings. Some iconic examples: In Paris the former railway station Gare d'Orsay, built in 1900, was transformed to one of the most attractive art museums in Paris, the *Musée d'Orsay* in 1986 (Figure 2.4). In London, it has been a problem to reuse the old *Battersea Power Plant*, built in 1929 and closed in 1983 (Figure 2.5). The Battersea Power Plant is a known characteristic landmark of London's skyline with four smokestacks, one in each corner. When it was built it became the largest brick building in Europe. Since the closure, many development plans have been presented without success. An example of reuse of a building and change to quite another purpose is the *Elbphilharmonie* in Hamburg. A former harbour warehouse built in 1923, it was retained at the base of the new cultural and residential complex. The building has a ground area of 120,000 square metres and a height of 108 metres (Figure 2.6).

In urban development, we often see dilemmas and problems of fitting old preserved buildings into new city surroundings. This is the case in many countries around the world, especially in Europe and the United States. If it is not possible to find an applicable purpose for the old buildings, the façades are kept and the rest of the building is demolished and integrated into new building structures, as show in Figure 2.7.

Reuse of special building structures, for instance reuse of high-rise silos for residential or commercial purposes is often seen in Europe. The major advantage of the transformation of a concrete grain silo to other purposes is the height of the structure. In city development, the height of new buildings is often restricted according to a political decision. Therefore, the approved height of existing industrial buildings, for example silos and towers, are exploited for other new purposes. In Denmark, many examples are seen with converted silos, where new openings are cut in the structures and a number of flats are constructed. A special design has been developed for reuse of a former chemical plant with a number of silos very close to the centre of Copenhagen. Figure 2.8 shows three reused silos, one has a rectangular shape and the two others are cylindrical. The rehabilitation design of the cylindrical silos is based on adding flats outside the original cylinder, which is used for lifts and staircases, utilities, and so on.

Figure 2.4 Musée d'Orsay, former railway station Gare d'Orsay in Paris.

2.5.3 Urban development

Multi-transformation of buildings in urban renewal takes place in all cities following the growth of population and the need for better housing. Besides the expansion of the cities and consumption of land, the most attractive development is the change of the old and outdated stock of industrial facilities and harbour areas into residential, cultural and commercial areas.

Development projects such as the *London Docklands* is a typical example of a big harbour and industrial development project. Between 1960 and 1980, all of London's docks were closed, leaving around 8 square miles (21 km²) of derelict land in East London. The planning of the development started in the 1970s. The massive development program was sped up

Figure 2.5 Battersea Power Plant in London. (Photo courtesy of Creative Commons.)

Figure 2.6 Elbphilharmonie in Hamburg. Rehabilitation and extension of old warehouse for residential and cultural purposes in 2016. (Photo courtesy of Ole Larsen.)

Figure 2.7 Old buildings in new environments. Examples from Washington, DC, and Brussels in 1991.

during the 1980s and 1990s, and a great part of the area was transformed into a mixture of residential, commercial and light industrial space. The development program included among others the *Canary Wharf* high-rise building, the highest building in London at that time; and a new local traffic airport, the *London City Airport*.

Another iconic harbour development program is the development of the Bilbao Harbour in Spain, where the area, after an economic crisis, was transformed to a cultural park including the internationally known Bilbao *Guggenheim Exposition*, designed by the reputable architect Frank Gehry.

Huge urban development projects are seen around the world. During the past decades we have seen great changes in the megacities of the Far East, for instance Shanghai, Singapore, Hong Kong, Kuala Lumpor and Jakarta. The transformation of Shanghai, whose 2013 skyline is shown in Figure 2.9, is very impressive. However, through the ages urban development has been performed more or less in a ruthless manner without specific consideration to the existing cultural and architectural values of the building stock (see for example, Brussels in Belgium, Figure 2.10).

Figure 2.8 Reuse of old silos for residential purposes in Copenhagen.

Usually, the multiple-building transformation is caused by economic development combined with growing population and the need for housing. The social political development is also a very important factor. Because of the lack of social development, poverty and poor design of housing, many examples of modern slums in urban development from the 1950s to 1980s are seen in European countries. A typical example is the *Ballymun Regeneration Project* from 2004 to 2017 in Dublin. The project comprised demolition of seven 15-storey blocks, nineteen 8-storey blocks, ten 4-storey

Figure 2.9 Shanghai 2013. (Photo courtesy of Creative Common.)

blocks and seven underpasses. The 36 residential blocks were built in the 1960s with prefab concrete elements. Because of social problems including drugs and crime, the Dublin City decided in 1997 to regenerate the residential area. It included a total floor area of 289,000 m² and a total amount of CDW assessed at 324,000 t (DEMEX 1998). The demolition of the first of the seven tower blocks took place in 2004. In 2015, the last tower block was demolished. Two of the buildings were demolished by explosives (implosion); the other five were demolished by mechanical ways with long-range concrete breakers. In Copenhagen, Denmark, a similar built-up area is planned to be renovated. Five out of twelve 16 floor, 50 m high buildings will be demolished in 2019 because of critical PCB pollution (see Figure 2.11). During the past years, the need for urban regeneration has accelerated because of the poor integration of immigrants combined with displaced persons and refugees from North Africa and the Middle East. The concentration of unemployed and social deprived ethnic groups calls for changes in former housing policies and urban regeneration like the Ballymun Regeneration Project from 2004 to 2017 as mentioned above. In Denmark, for instance, the government has addressed requirements for the regeneration of a number of poor housing communities called *ghetto areas*.

The demolition of the *Kowloon Walled City* in Hong Kong, in 1993, was a political issue. Ever since the United Kingdom leased the New Territories on the Chinese mainland in 1898 – for a period of 99 years, the Kowloon Walled City had been a political probleme area. According to an agreement between the United Kingdom and China the Walled City should have been

Figure 2.10 Example of implementation of an urban renewal plan in Brussels, 1991.

demolished before the Chinese takeover of Hong Kong in 1997. The Walled City housed about 34,000 people living in approximately 500 buildings, composed of 12–14 storey tower blocks, within an area of approximately 20,000 m². The demolition project was planned and managed by the Civil Engineering Department of Hong Kong. The demolition contract, worth approximately $7 USD million, was performed by a joint venture between the local contractor Express Builders Co. Ltd and the US firm Cleveland Wrecking Co. The main demolition method was crane and ball, which was cheaper than demolition by explosives. Approximately 170,000 m³ of CDW was crushed to sizes below 260 × 250 mm and dumped at government-selected designated areas (see Figure 2.12). After demolition, a public park area was established on the site (Lauritzen 1993).

Because of urban development and increased demand for space many city airports have moved to new areas, and left the old areas for development of new housing. This was the case for the Fornebu Airport in Oslo, Norway, and Kai Tak Airport in Hong Kong.

Great international events like the Olympic Games and world expositions require huge amount of buildings and facilities. The Olympic Games in London 2012 took place on a 2.5 km² site, Queen Elizabeth Olympic Park.

Figure 2.11 Five tower blocks, 16 floors, 50 m high, of the Brøndby Strand buildings in Copenhagen are planned to be demolished in 2019 because of PCB pollution beyond repair.

The Olympic Park project involved the demolition of over 215 buildings alongside a number of walls, bridges and roads. A recycling rate of 98.5% (427,531 tonnes) was achieved (Carris 2014; Deloitte et al. 2012). More details of this multiple-transformation project are presented in Chapter 5, Section 5.4.

2.5.4 New buildings with recycled materials

Through the ages, it has been common practice to save and reuse building materials. Such as bricks, tiles, wood and steel. Today, we are more restrictive and reluctant to use recycled materials in the construction of new buildings. If building a new structure with recycled materials, the structure must meet the requirements of new materials, for example according to the EU harmonised condition for the marketing of construction products (EU Parliament Regulation 2011). For some materials, standards might exist. This is the case for the production of new concrete with recycled aggregate in Europe and the United States. However, one needs to convince the building owner, his investors and assurance companies that the recycled materials substitute natural materials without any risk to the quality or the value of the structure.

Figure 2.12 Demolition of the Kowloon Walled City. 34,000 people lived in the 20,000 m²
area.

The key issues of building with recycled materials are:

- To prove the cost-benefit of building the specific structure with recycled materials
- To document the quality in accordance to standards of new materials or standards for recycled materials
- To establish the logistic match between the need for the materials in the new construction and the availability of recycled materials

These issues, as well as examples of new buildings constructed by recycled materials, arc presented and discussed in detail in Chapters 4 and 5.

2.6 HAZARDOUS MATERIALS, THE MAJOR CHALLENGE

One of the major problems with recycled building materials is the issue of toxic substances in the existing buildings and structures, as indicated by the Ellen MacArthur Foundation in Table 2.3. From ancient times, we know copper, lead and mercury were used in building materials. In the nineteenth century, a number of toxic materials were used of various reasons, such as hardening, plasticizing, fire protection and impregnation. In the period from the 1950s until the 1980s, when the impact of toxic substances in building materials became known and their use stopped, asbestos and Polychlorinated biphenyl (PCB) were employed extensively.

The industrial world continues to use large quantities of chemical substances in all types of products in order to enhance technical properties, extend a product's life, and so on. Some of these substances have already been shown to have an impact on the environment and/or human health. Therefore, the authorities in EU and other European countries have enacted regulations or bans to counter these impacts. Denmark and The Netherlands have made extensive research on harmful substances in CDW in general. The EU, Norway and Sweden have made focused studies for example on PCB in open use in the construction industry. The environmental impact of asbestos fibres and their countermeasures have been known for many years, and the asbestos handling is regulated by well-defined actions. However, the asbestos handling must still be considered as an important activity in the CDW management since the problem exists.

Referring to Danish experiences and information collected during the IRMA Project (IRMA 2006), the following 12 harmful substances – besides asbestos – have been identified as the most problematic substances in building waste materials:

- Lead
- Cadmium
- Mercury
- Nickel
- Chromium
- Copper
- Zinc
- Polychlorinated biphenyls (PCBs)
- Chlorinated paraffin
- Chlorofluorocarbons (CFCs)
- Hydrochlorofluorocarbons (HCFCs) and hydrofluorocarbons (HFCs)
- Sulphur hexafluoride

These substances were selected from a long list of undesirable substances. The selection was based mainly on toxicity (either on the environment or on human health) and the consumption of the substances. An inventory of the 12 substances – *The Dirty Dozen* – was made, describing their application, consumption, toxicity and anticipated disposal processes for the waste containing them.

2.6.1 Asbestos

Asbestos is a carcinogenic material. Asbestos fibres that are inhaled by humans may cause cancer and asbestoses. Such health effects have occurred many times in the past years, especially in cases of employees at asbestos plants. Therefore, asbestos is well known, and generally scary to the public. In relation to these health effects, asbestos is probably the most problematic substance in buildings that are refurbished or demolished. The use of asbestos is prohibited, but because of its broad applications in the past and its durability, asbestos is still present in many buildings today. The risk depends on the way asbestos is found. Free fibres pose the greatest risk for health and safety. Cement bound asbestos present the lowest risk, if not weathered.

The removal of asbestos increases the refurbishment and demolition costs significantly. In practice, the removal of asbestos has therefore led to health and safety problems, because the removal was not performed thoroughly enough, even though a certificate was obtained stating that the building was free of asbestos. With respect to the removal of asbestos, a more robust and practical working program needs to be developed in order to deal with the asbestos problem.

2.6.2 Organic substances

Organic substances might cause health problems and other environmental problems. Furthermore, they might be problematic for reuse and recycling. Examples of problematic organic substances are poly-aromatic hydrocarbons (PAHs) (from tar) phenols and toluene (from glues), plasticizers (from synthetic materials), PCBs (from plasticizers, painted surfaces, sealing materials transformers and condensators, as cooling liquid or lubricants). PAHs and PCBs are carcinogenic. Other organic substances might cause health problems in the long run if a worker is often exposed to them, such as volatile organic substances (VOS) in for instance oil and solvents (paints, glues). Direct health problems seldom occur, except for inconvenience like smell or irritation of eyes. Furthermore, pesticides can be found in building materials (front walls) to prevent them from becoming mouldy, as well as halogen compounds which are used in coatings to increase fire resistance.

2.6.3 Heavy metals

Heavy metals might be toxic for humans. The risk involved depends on the mobility of the ions or compounds. A high content does not always imply a high risk. As a result of negative health effects on workers, lead is a prohibited substance in paint nowadays. Nevertheless, during the next 10–20 years, lead in paint will remain a problem for the working environment during refurbishment and demolition.

Leaching of heavy metals is an environmental problem, which can occur for instance when recycled materials are applied in new materials. In order to prevent the subsoil from being contaminated when potentially contaminating materials are used in constructions, national decrees have been enacted, for example the Dutch Building Materials Decree gives leaching requirements for heavy metals (As, Ba, Cd, Co, Cr, Cu, Hg, Mo,

Ni, Pb, Sb, Se, Sn, V and Zn). These heavy metals have been identified as most common in building materials.

It should be noted that the requirements of the national standards, generally, only apply to mineral materials. Leaching of heavy metals from paint, zinc (gutters, etc.) and plastics for example are not subject of this legislation.

2.6.4 Other contaminants

Besides the above-mentioned ones, specific contaminants can be found in industrial constructions, depending on the type of industry. Other contaminants to be handled with respect to the health and safety of demolition workers and/or environmental impact are the following;

- *Quartz*: Quartz dust is a fine mineral dust. When inhaled by humans it may cause lung diseases
- *Fibres other than asbestos*: Fibres in mineral wool and other materials get attention in relation to health and safety, especially since the asbestos situation
- *Foams*: Foams used mainly for insulation contain foaming agents, which may be released during demolition. Most foaming agents entail no health risk, but might be problematic for environmental reasons
- *Radon*: Radon is a radioactive gas, which is released from building materials
- *Nanomaterials*: The very small nanoparticles are health critical like asbestos materials. The future use of nanomaterials must be controlled

2.6.5 Need for cleansing materials

The risk of toxic materials in recycled materials is hampering any initiatives for recycling and substituting natural/primary resources with recycled/secondary resources. Therefore, it is mandatory to assess the type of pollution and identify the toxic materials before demolition and the planning of recycling.

Chapter 3

Demolition

On the seventh day you shall go round the city seven times. Then you will blow a long tone on the horns, and everyone will raise a huge war cry. Then the walls will collapse!

Demolition of the Walls of Jericho, Joshuás Book 6.1–20

3.1 INTRODUCTION

Demolition is the general term for a set of actions to remove a structure, building, part of a structure or a building. Depending on the type of structure we might use another term. For example: decommissioning of nuclear facilities, dismantling of offshore structures and deconstruction of special structures or deconstruction of buildings.

A typical demolition job starts with a need for changing a structure. The building owner plans a demolition project depending on the type and size of the job. In case of urban area renewal, removal of several buildings, big and tall buildings or complex structures, the demolition needs planning and project management. In case of a small house or structure, the building owners hires a skilled demolition contractor to do the job.

The demolition project starts with planning including pre-demolition audit, assessment of the building and design for demolition followed by the tendering process and contract. Usually, the demolition contract starts with preparation of demolition including the establishment of the worksite, cutting of and changes of utilities (telephone, electricity, sewage, water, municipal heating, etc.). Left items and garbage are removed and the building is cleansed outside and inside. All toxic and environmental harmful substances such as asbestos, organic substances, heavy metals, and so forth, are removed. Demolition is carried out either as *total demolition* or as *partial demolition*. Total demolition involves complete removal of buildings or structures. With partial demolition only parts of a structure are removed, for instance in case of rehabilitation and repair projects. The best possible basis for recycling of waste is achieved by *selective*

Figure 3.1 Typical elements of demolition planning, cleansing, demolition and CDW management.

demolition, because it facilitates the sorting of CDW at the source. By selective demolition is understood a process in which demolition takes place as a reverse construction process where the different types and fractions of materials are removed from the building and sorted so that mixing of bricks, concrete, wood, paper, plastics and other materials is avoided (see Section 3.3). Sorting and handling the CDW takes place in accordance to the national legislation and local rules for recycling, waste disposal or other treatment. Sustainable CDW management should be based on the requirements of resource efficiency and green change in accordance to the UN Sustainable Development Goals 2030, the Paris Climate Agreement and circular economy, presented in Chapter 1.

This chapter gives an overview of the activities and the framework for the demolition activities, presented in Figure 3.1, followed by a brief presentation of cleansing technologies and demolition technologies. At the end of this chapter, methodologies of demolition and implementation of the technologies are presented with practical examples.

3.2 PLANNING OF DEMOLITION PROJECTS

Demolition is a work for specialists with special knowledge of old structures and how to remove them. In former times, detailed planning of demolition projects were more or less ignored, because it was up to the demolition contractor to plan and implement his job within the agreed time and cost. Today, demolition contracts are subjected to the same conditions as for all other contracts in the building and construction industry meeting the requirements of legislation and technical specifications with a special focus on health and safety.

The planning of demolition projects comprises:

- Preliminary investigation, including a historical review, screening of contamination and materials, assessment of environmental constraints and initial estimation of time and budget
- Pre-demolition audit, including assessment of amount and types of CDW and resources, assessment of contamination of the demolition object

- Design for demolition, including assessment of work methodologies, change of infrastructure, environmental management, health and safety management, information management
- Tendering, including tender documents, procurement and contract

3.2.1 Preliminary investigation

The preliminary investigation comprises assessment of the size of the demolition project, specific problems, constraints and draft budget for the demolition work. Based on existing drawings and inspection of the sites and buildings, as well as the total ground and floor areas, the main fractions of building waste and the total amount of waste are assessed. In most cases, especially for old buildings, no drawings exist. Aerial photos and Google Earth can roughly provide measures of the buildings to be demolished. The size of the demolition projects in terms of masses, volume, ground area and floor area is roughly estimated. Useful figures for the estimation of weight and fractions of buildings are presented in Tables 3.1 and 3.2. The figures are derived upon experiences in Denmark but might be used in other countries with similar types of buildings.

3.2.2 Historical review

A review of the history of the building and the use of the building gives the first indication of the range and the type of possible pollutions such as asbestos, oil, organic substances and heavy metals. Old buildings, specifically dwelling houses, constructed before the 1950s, are not expected to contain

Table 3.1 Example of unit-masses of waste from the construction of new buildings, construction and demolition waste from renovation of existing buildings, and waste from demolition of buildings

	Construction		Renovation		Demolition			
	Housing/ Commercial		Housing/ Commercial		Housing		Commercial	
Type of CDW	kg/m²	Pct.%	kg/m²	Pct.%	kg/m²	Pct.%	kg/m²	Pct.%
Concrete, tile	18	78	31	62	1,510	93	1,400	80
Wood, inflammable	3.5	15	13	26	110	7	90	5
Paper, cardboard	0.5	2	–	–	5	<1	–	–
Plastic	1	5	–	–	–	–	–	–
Metal	–	–	5	10	–	–	90	5
Other, incombustible	–	–	1	2	–	–	180	10
Total	23	100	50	100	1,625	100	1,760	100

Source: Lauritzen, E. K. and Jacobsen, J. B., Nedrivning og Genanvendelse af bygninger og anlægskonstruktioner, SBI Anvisning nr. 171, 1991 [In Danish].

Note: The square metre is floorage, in average 3–4 floors.

Table 3.2 Example of content of mapping contamination of a structure

List of content	Comments
1. Purpose of the mapping	Clear description of the objective, e.g. initial planning, authority application, etc.
2. Limitations and preconditions	Partly or full assessment, focus on specific pollutants and issues
3. Available building information	Missing information, uncertainties
4. Performance of sampling	Description of following: • Consideration on numbers and places of samples • Surface samples and/or drill core samples • Unambiguous identifications and marking of samples • Used test equipment, packing and transport of samples to analysis laboratory
5. Photos and drawings	Photos of sampling place and other relevant information clearly explained
6. Analysis	• Description of laboratory and accreditation • Description of analysis methods, instruments and standards • Report of results including limits of detection, references of threshold values, uncertainties, etc.
7. Report	Includes: • Introduction and general view of samplings and results • Uncertainties • Information and interpretation on the use of the report • Conclusions • Recommendations

Source: Danish Competence Centre on Waste and Resources, DAKOFA, 2017.

hazardous materials in the same degree as buildings constructed from the 1950s to the 1980s. However, if the buildings have been rehabilitated, repaired or maintained after the 1950s they might be polluted. Buildings constructed between the 1950s and the 1980s might contain all kinds of hazardous materials. In Denmark, it is mandatory to screen all buildings and report to the municipal authorities about the risk of PCBs in structures built between 1955 and 1982. Regarding older industrial buildings and facilities, from the time before environmental protection and pollution control, the probability of pollution with oil and chemicals might be considerable.

3.2.3 Screening of contamination

Depending on the first inspection and the historical review, an initial screening is conducted of the building for possible contamination. The objective of the screening is to identify the types of pollution and the scale of the needed decontamination work with respect to time and cost. The screening for pollution forms the basis of the detailed assessment and mapping of contamination.

3.2.4 Screening of materials

It is recommended at an early stage of the project to establish an overview of the quantities of the waste materials and the opportunities for recycling. The amount of machinery, steel structures, steel in reinforced concrete and other kinds of steel and metal might influence the cost of demolition with positive income. In case of a big demolition project, for instance more than 10,000 t of waste and mainly concrete, it is recommended to make an initial assessment of the various opportunities for recycling.

When planning the demolition and CDW management, it is economically and technically advantageous to separate the waste at a location as close as possible to the source where it arises (selective demolition). In principle, this implies that the demolition contractor should sort all fractions of waste on-site. Mixing of waste and later sorting at a different location is usually more costly. Moreover, it can be associated with environmental problems concerning dust, noise and smell as well as health and safety hazards.

3.2.5 Assessment of environmental constraints

The site and neighbourhoods around the demolition site must be assessed with respect to risk of damage to third-party property, or inconveniences to the neighbours and public. The assessment comprises the expected impact of the demolition processes with respect to vibration, dust, noise, odour, traffic inconveniences, and so on.

3.2.6 Initial estimation of time and budget

Based on the physical size, history, scanning of pollution and environmental constraints, the needed time and budget of the project is presented to the building owner. Because of financial reasons, it is often the case that total demolition of obsolete buildings is carried out immediately before the start of a new building project. Therefore, the time schedule for demolition work is usually so tight that there is very little time for measures to protect the environment and to prepare for the recycling of CDW. This is due to the fact that the building owner often ignores the priority and importance of the demolition work and CDW management.

Cost estimation of the demolition contract is difficult and very uncertain. In general, demolition contractors work locally/regionally and the demolition costs depend on the local conditions and market. The demolition costs comprise four principal cost elements:

- Preparatory works, including establishing worksite, stop and changes of utilities, etc.
- Cleansing of buildings and removal of hazardous substances including handling critical and hazardous waste
- Demolition work including selective demolition, health, safety and environmental management
- CDW management and handling including recycling of CDW and sale of scrap materials

Often, the cleaning costs and the CDW handling costs are higher than the demolition cost.

3.2.7 Pre-demolition audit

Referring to the EU directives on implementation of a circular economy, presented in Chapter 1, it is expected that the building owner must assess the materials of the buildings before demolition and prepare a plan for recycling the materials. Based on existing drawings of the building and visual inspection, all types and the amounts of materials are mapped systematically. For each fraction, opportunities for reuse, recycling, recovering or other kind of utilization are estimated, for instance reuse of bricks, recycling of crushed concrete, recovery of scrap metal, and so forth. The European Commission has introduced *pre-demolition audit* as a term of assessment of resources, which has been adopted as a mandatory instrument in many member countries.

3.2.8 Assessment of CDW and resources

Today, the management of CDW is a part of the demolition contract, and the building owner does not pay much attention to this part of the contract. Normally, the demolition contractor sells the scrap metal or enter a subcontract with a scrap dealer. After decontamination, the inert materials such as concrete and masonry are transported to recycling companies or landfill areas. In case of a bigger demolition project with a high volume of concrete, the demolition contractor requires permits for the establishment of a crusher on-site to crush and reuse concrete materials for another construction site.

The planning of the building owner comprises a review of his obligation according to national and local rules for waste management. In many EU countries, for instance, in Denmark, it is required that the building owner assesses and reports to the municipal authorities the types and amounts of waste in the respective fractions, for example: inert waste (concrete and masonry), organic waste (wood), combustible waste (paper, plastic, textile, etc.) and hazardous waste (asbestos, organic hazardous waste, heavy metals).

In case of big demolition projects, for example urban renewal projects, power plants, offshore facilities, production facilities, bridges, airports and so forth it is recommended that the building owner assesses the opportunities for recycling the materials and identify a potential buyer for the recycled materials. Then, the building owner can establish a budget of waste management as a basis for the tendering and discussion with the demolition contractors.

3.2.9 Assessment of contamination

Based on the initial screening and identification of possible contamination of the building, a detailed inspection of the building takes place in order to provide a complete assessment and mapping of the type and extent of contamination inside and outside the building.

The contaminant originates form the building materials, building processes or the use of the building. In unsuspected buildings (e.g. dwellings and office buildings) most of the chemical contaminants can be found in coatings, joints, glues and specific parts like chimneys and roofs. Physical contaminants (asbestos) can be found throughout the entire building in various places (roofs, walls, pipes, coatings, etc.). In suspected buildings (e.g. industrial buildings), the contaminants are also related to the processes that have taken place there.

Knowledge of the history of the building can give sufficient background information in order to identify the types of contamination. The most widespread technique is selective or environmental demolition in which known contaminations are removed before the carcass of the building is completely demolished. Considering demolition, it is also necessary to look at the recycling possibilities of the demolition materials. This can influence the demolition process.

The possible polluted areas and constructions are mapped and marked. Specimens of the structures and surfaces are taken for laboratory analyses. Based on the results of the laboratory analyses, a detailed report on the specific pollutions of the buildings must be prepared (see examples in Table 3.2).

In general, it is recommended that specialized and independent consultants or laboratories perform the contamination assessment in order to ensure a qualified and sufficient assessment, which is not affected by the economic interests of the involved partners.

Many building owners prefer to include the detailed investigations of the contamination of the buildings in the demolition contract, hoping to get a cheap price and to avoid any extra bills for unforeseen/hidden pollution. Unless specific rules and standards exist for the assessment, the contactor's assessment might turn the tendering into a competition for the easiest and cheapest decontamination work. The winner is the contractor offering the cheapest price for the decontamination – he might have found less harmful substances than the other bidders and not conducted specimen and laboratory analyses. The loser is the contractor that takes the responsibility of environmental protection seriously. Often, it will be difficult to control the quality of the contractor-designed decontamination work.

The need for material tests, including specimens taken on the site and laboratory analyses, is always an issue of discussion. Depending on the type of pollution and the methodology of the analyses, the cost of testing must be considered carefully. Therefore, the estimated number and types of tests must be based on expert assistance.

3.2.10 Design for demolition

The design for demolition shall meet the requirements of cleansing, safety requirements, environmental requirements and specific functional requirements. The design should comprise the following elements besides the pre-demolition audit:

- Assessment of work methodologies
- Change of infrastructure
- CDW management
- Environmental management
- Health and safety
- Information management

3.2.10.1 Assessment of work methodologies

During tendering a demolition contract, the building owner might see the advantage of giving the bidding contractor free choice of technologies and project design. However, with respect to quality, risk and safety together with the requirements of environmental projection, it is generally recommended that the building owner should take the responsibility for the overall project design based on specific requirements. This will also contribute to an open and equal basis for the tendering and competition. Then, the appropriate details and interfaces of the design can be discussed.

A typical example is the design for the demolition of a high-rise structure, for instance buildings, towers and chimneys. The building owner will consider the various opportunities for demolition by blasting, the use of crane and ball, long-arm concrete crushers, or top-down demolition using small machines. On this basis, he will prefer one or more specific solution and reject the others. The demolition contractor designs the details of the methodology. For the building owner's selection or rejection of methods for a critical demolition, a risk-based approach is recommended (see Section 3.7).

3.2.10.2 Change of infrastructure

Most demolition jobs require cutting or changes of the feed circuit (water, heat, gas, telephone and Internet) and outlet (drainage of surface water and sewers in accordance to local authorities' requirements and the respective owner's guidelines). In some cases, there is a need for greater changes of the circuits and the sewers. Usually, this kind of work is performed as a separate contract before the start of the demolition contract. It is important that the infrastructure project is coordinated with the demolition contract in order to avoid delays of the work.

3.2.10.3 CDW management

Referring to the pre-demolition assessment, CDW management must be integrated in the overall demolition design. The requirements of cleansing and handling CDW together with responsibilities of the owner and the contractor must be clearly specified.

3.2.10.4 Environmental management

The building owner is responsible to the neighbours with respect to following the requirements of environmental protection. Depending on the national and local rules, he/she might delegate some of the responsibility to the demolition contractor. However, the building owner must review and assess the risk of environmental impact comprising:

- Risk of vibration impact on sensitive buildings, installations, computers, etc., including inspection of buildings and photo monitoring of existing cracks and damages
- Risk of noise exceeding the public threshold levels
- Risk of dust damages to sensitive installations, hospitals, etc.
- Inconveniences of the normal traffic in the area
- Risk of rats

Much experience of monitoring and control of noise and vibration exists. Control of dust might be critical, especially during blasting operations. Critical wind directions and sensitive neighbours must be assessed, and precautions to avoid dust problems must be planned. Special attention should be paid to air intake of hospitals, clinics, sensitive production plants, and so on in the surrounding area.

Opportunities for control of dust, noise and vibration must be investigated and the transport of materials from the site must be planned and coordinated with the local authorities and police.

3.2.10.5 Health and safety management

The worker's health and safety is an important issue. Demolition work is known as high-risk work because of the exposure of the workers to the special work conditions with respect to awkward work positions, stabilisation of structures during the demolition processes, and so forth. Generally it is recommended to keep workers away from the structure. This means that long-range machines and robot tools are preferred over disassembly and demolition by hand. However, this might cause a dilemma with respect to carful dismantling of materials by hand for reuse.

It is expected that the building owner prepares a health and safety operational plan or standard operational procedure for the specific demolition job. Furthermore, the building owner must organize a health and safety organization clarifying the responsibilities of the safety and health procedures and the current monitoring of the work.

3.2.10.6 Information management

Big and spectacular or politically sensitive demolition projects are often a subject of comments and attract the interest of the press. Therefore, it is recommended that the building owner consider the publicity and the risk of negative publicity. Depending on the importance of the demolition project, the building owner should take the necessary steps for implementing a media strategy in order to avoid any bad publicity.

3.2.11 Tendering

Generally, tendering, procurement and contract of demolition projects follow rules and guidelines like other projects within the building and construction industry. The rules for EU countries on public procurement are stated in the EU Directive of 26 February 2014 on public procurement (EU 2014). Depending on the size of the contract, the tendering process often starts with prequalification followed by shortlisting of four to six companies or consortiums. The pre-qualification application requires information on the experience and capacity of the company with respect to the specific job.

The tender documents, which varies from country to country, comprises typically:

- Letter of invitation to bid, including requirements of the bid
- Scope of work
- Work description
- Condition
- Financial offer

Often, a two-envelope process is used. The first envelope contains the technical proposal, which must be evaluated and ranged before opening the second envelope with the financial proposal. The purpose with the two-envelope process is to ensure an evaluation of the technical proposal without influence from the financial proposal.

Referring to Figures 3.1 and 3.2, it is recommended that the time schedule should not stress the tendering process and implementation of the demolition. In order to receive the most advantageous bids, the bidders must have time to investigate opportunities for the cheapest solutions for

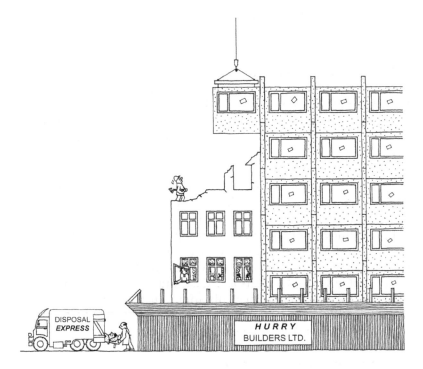

Figure 3.2 Time schedules for demolition work are usually so tight that it is not possible to carry out demolition in an optimal way from an environmental protection and recycling CDW point of view. (According to Lauritzen and Jakobsen 1991.)

CDW management, including opportunities for recycling of CDW. New types of tendering procedures based on dialogue might be appropriate for demolition projects focusing on recycling CDW. Compared to the work description of a construction project, the demolition work is described in lesser details because the work processes are not standardized to the same extent. Often, there is a discussion of the extent and details of the description of the demolition work. To avoid misunderstandings or discrepancies with the information of the work, some building owners prefer to minimize the information and description to the most basic information of the buildings to be demolished. Following this point of view, the bidders' technical and financial proposals are based on their own calculation of the job parameters, such as ground floor area, total floor area, masses and fractions of CDW, and so on. This leads to a competition based on the bidders own estimates and risk of miss-calculations. In principle, all relevant information on the building and the work should be disclosed. In case of a dispute between the building owner and the contractor, the building owner might be sued if he/she has retained relevant information.

To ensure the highest quality of the proposals and an equal basis of the proposal, it is recommended that the work description comprise the following:

- Detailed description of the decontamination work with the contamination assessment report enclosed
- Description of the buildings including historical description, type of buildings and constructions, estimation of floor area, eventually also masses of CDW and all available drawings and documents related to the buildings
- Instructions and requirements of demolition methods, preferred demolition methods, rejected demolition methods or the contractor's own choice, optimization of demolition methods favouring recycling of CDW
- Instruction and requirements of CDW management, including opportunities for recycling, ownership of materials left in the building and the CDW, ownership of recycled materials, etc.
- Instruction and regulation of environmental management, including acceptable level of vibrations, dust and noise, requirements of the contractors for monitoring vibration, dust and noise, with special regard to the traffic, intermediate change of traffic and other measures to protect the surrounding environment and neighbours
- Instructions and requirements of health and safety including risk management and safety organisation
- Requirements of quality management, quality assurance and quality control

In order to avoid misunderstanding of the estimated figures and to eliminate discrepancies in the work description, the building owner and the contractor should discuss and agree on the figures before the start of the work.

A demolition work is characterized by many uncertainties related to the description of the demolition job, possible changes in the work and extra work must be foreseen, for instance in case of hidden or unexpected polluted constructions, or when the constructions are not in accordance to the drawings, and so on. Therefore, it is very important that procedures regarding the contractor's claims on extra or changed work are described and agreed to. Usually, the demolition work will start immediately after the contract signing with the preparation of the worksite.

3.3 DEMOLITION WORK

All demolition contracts comprise the following typical activities:

- Preparatory works including change of infrastructure and establishment of worksite
- Cleansing of contaminated parts of the building

- Selective demolition
- Sorting of CDW
- Finishing, including clearing of worksite and finishing

Principles of cleansing and demolition processes are described in a general context in this section. The cleansing technologies are presented in Section 3.5. Demolition methodologies and technologies are presented in Section 3.6.

3.3.1 Preparatory works

3.3.1.1 Change of infrastructure

After end-of-use and decision of demolition the building must be prepared for demolition. Depending on the ownership of the building and the transformation process the infrastructure needs to be changed. All utilities, such as sewages, heating, water supply, telecom, and so forth, must be cut off, or changed in accordance to the plan for the future use of the site with the approval of local authorities. In case of major changes, the work might be carried out as sub-contracts or contracts outside the demolition contract.

3.3.1.2 Worksites

The worksite is established and fenced around the building. The worksite must include a site office, shantytown work sheds, space for the containers and depots for the various fractions of CDW after sorting, and suitable working space around the building. The worksite management must prepare the necessary means to protect the neighbours against environmental impact from the demolition work. In order to control dust, the building should be fully shielded. The demolition equipment and machines must meet local requirements of energy consumption, noise and CO_2 emission. If necessary, systems for monitoring vibration, noise and dust should be established outside the worksite.

3.3.1.3 Removal of garbage and left items

Usually, the demolition contractor takes over ownership of all items left, and it is his responsibility to remove and handle all items including garbage. Special attention should be paid to human waste and needles left by drug takers who used the empty building before demolition. Also, control and extermination of rats can be an important issue.

3.3.2 Cleansing

By the start of the contract, after the establishment of the worksite, the cleansing of buildings commences in accordance to the assessment and mapping of harmful substances (see Chapter 2, Section 2.6) and the work description. The cleansing work aims to clean all contaminated materials and surfaces outside as well as inside. Some contaminated items and materials, for instance pipes insulated with asbestos, are removed for handling and cleansing outside the building. Contaminated surfaces are treated by the use of various technologies: mechanical, mechanical-hydraulic, and milling (see Section 3.5). In order to control dust emission from cleansing work, airtight work places are established. In some cases, it might be appropriate to seal the entire building to control the risk of hazardous dust emission.

3.3.3 Selective demolition

During traditional demolition, it is the normal work method to crush the building from the rooftop to the basement without specific consideration on sorting the waste. Selective demolition aims to sort the waste on-site in order to exploit all opportunities for recycling CDW. Selective demolition comprises typically four stages, which are presented in Table 3.3.

Stage I is the removal of all objects that are not part of the building including abandoned furniture, paper, garbage, and so on. The removal work depends on how long the building has been left without monitoring and if it has been accessible to intruders of all kinds.

Stage II is the removal (stripping) of installations apart from the main bearing structure, including: roof, windows, doors, floors, installations, tiles, partitions, wall covers, plaster, insulation, and so forth. The items are removed and sorted into the various fractions (see Table 3.4). It is important that all organic and ceramic materials and other materials that cause problems to recycling of concrete are removed.

Table 3.3 Stages and activities of selective demolition

Stages	Activities
I. Preliminary work	Removal of remaining furniture and equipment
II. Removal of permanent Installations (stripping)	Removal of: • Doors, windows and frames, kitchens, toilets, etc. • Light partition walls, floors • Pipes and electrical installations • Asphalt on floors and tiles on walls • Roof
III. Demolition of shell structure	Upper parts by manual labour Lower parts by machines
IV. Removal of containers and general cleanup	Transport and delivery to recycling plants for waste materials, which to the largest possible extent are sorted into clean fractions. Transport and delivery of non-recyclable waste to incineration plants or controlled disposal sites

Table 3.4 List of fractions according to the European Waste Code (EWC 17
Construction and Demolition Waste) and typical handling processes

EWC Code 17	Waste materials fractions	Building member	Typical handling process
1701	Concrete, bricks, tiles and ceramics	Foundation, bearing structure, internal walls and slabs	Crushing and recycling, substituting natural materials
1702	Wood, glass and plastics	Roof, floors, doors and windows, membranes, covering	Reuse of wood, doors and windows, recovering of glass and plastics
1703	Bituminous mixtures, coal tar and tarred products	Roof and surfaces	Recovering, special treatment
1704	Metals (including their alloys)	Structural steel in bearing structure, reinforcement, pipes, installations, etc.	Recovering of metal and iron and other not hazardous metals
1705	Soil (including excavating soil from contaminated sites), stones and dredging soil	Foundations and filling around the structure	Reuse of clean soil, cleansing and recovery of soil containing hazardous substances
1706	Insulation materials and asbestos-containing construction materials	Insulation in façades, roof, heating and cooling installations, etc.	Reuse and recovering of not-hazardous materials. Special treatment of hazardous materials, disposal of asbestos materials
1708	Gypsum-based construction material	Partition walls, covering plates	Reuse of plates, recovery of gypsum
1709	Other construction and demolition wastes	All kind of structures, painted materials, etc., containing hazardous substances	Cleansing, recovering, special treatment and disposal

When the stripping is completed, the main structure including walls, decks, bearing construction and basement will be demolished and handled in clean recyclable fractions of concrete, masonry, stones and steel. The choice of demolition technology methodology depends on the height of the building, thickness of walls, and so on (see Sections 3.6 and 3.7).

Based on the results of Danish demonstration projects, it is estimated that selective demolition and sorting requires approximately 30% more labour than conventional demolition and dumping, but taking into account the magnitude of disposal fees charged in connection with traditional demolition it turned out that selective demolition in these particular demonstration projects was 17.5% less expensive than conventional demolition. However, it was concluded that selective demolition could give rise to increased health and safety hazards because of the additional manual labour involved in the sorting of materials on-site. On the other hand, it was concluded that most of the risks could be avoided by proper planning of the work and by use of special tools.

3.3.3.1 Manual demolition

In case of very small demolition jobs, lack of space or lack of mechanical equipment the demolition is carried out by handhold tools, such as hammers, beakers, handheld pneumatic tools, electrical tools, and so on. Manual demolition is typically used in cases of rehabilitation and repair of buildings and structures. It involves also the use of small machinery for the purpose of loading and removal of debris. The hand tools are operated and carried by the workers and may be powered by electricity, as well as hydraulic or pneumatic pumps. This procedure is the only one applicable in many places, especially for partial demolition, or where there is not enough space or supporting capacity in the structure.

Because of such reasons, despite widespread mechanization, it is the most widely-used technique to execute demolition works, although having a higher cost compared with others systems. Manual demolitions are in many cases used as an additional procedure to avoid damages in remaining parts of the building or adjacent buildings, and as a preparation prior to mechanical demolition or blasting demolition, and so on. Because of the work time, health and environmental constraints, such as the risk of *white finger disease* caused by vibrating pneumatic/hydraulic hammers, manual demolition is expensive and should be kept to a minimum.

3.3.3.2 Mechanical demolition

The demolition industry has developed a large number of types and sizes of tools for all kinds of mechanical demolition. Important drivers for these technical developments are workers' health and safety, protection of the environment by reduction of emissions of noise, dust, vibration and CO_2 together with reduced energy consumption and improved productivity.

3.3.3.3 Light equipment

Mechanical equipment less than 3,000 kg of weight is considered as light equipment, for instance small hammers, concrete crushers, backhoes and excavators. Also diamond cutting is light equipment. For choosing such equipment several criteria have to be considered such as the weight of the equipment, its dimensions, mobility, attachments available and visibility. Many other considerations have to be made, but a first election has to be done before to consider the practical aspects of the work to do. The use of these types of equipment is growing and manufacturers are introducing numerous models to the market to cover a wide range of demolition works. The materials to be worked on by light equipment may be masonry or concrete, but always keeping in minds the limited power of these tools. Light equipment is often used for interior demolition and top-down demolition of buildings.

3.3.3.4 Medium equipment

This includes equipment having a weight between 3 and 30 tonnes and is the typical size of machines for demolition work in general. Because of its greater size, it is possible to use attachments (hammers and shears) of a larger size and power. Normally, the machines are used at ground level or on very strong slabs and work platforms, usually industrial building with a very strong structure. It is easy to transport this equipment, but it has a limit due to its reduced reach range.

The medium size excavator with a bucket, often a backhoe bucket, is the most commonly used demolition machine. Demolitions of family houses built of masonry and minor one- to three-storey buildings of masonry, or concrete elements, are carried out with a typical medium 20–30 t weight excavators mounted with 1.5–2 m³ bucket, which can reach buildings up to 4–6 metres high, enabling both demolition and loading of the building materials on trucks. Excavators mounted with applicable types of buckets enable the operator to sort out the materials during the demolition process. The bucket can be replaced with a hydraulic hammer, which improves the flexibility of the machine.

3.3.3.5 Heavy equipment

Equipment such as excavators with buckets, hammers, concrete crushers and long-reach booms (long-arm), weighing 30 to 100 tonnes or more is considered as heavy equipment. It can be used for cutting of concrete or steel structures. The performance of the heavy equipment has been improved over the past years. Manufacturers are developing a very specialized range of booms to achieve the maximum distance, while introducing better materials and hydraulic cylinders. Such distances need a better dust reduction system, provided by the use of water supplying circuits in the boom. Also, the distance from the operator to the attachments requires a video monitoring system and a high degree of training by the operator. A variety of buckets and grabs for all kinds of purposes have been developed. The long range booms and video monitoring on the tools have enabled the operators to perform very precise and careful demolition operations. Today, ground-based equipment can reach structures with a height of up to 50 m and in some cases even higher. However, this kind of equipment is very specialized and expensive.

3.3.3.6 Sorting CDW

The materials from the demolition are collected on the site in fractions and loaded in marked containers. A suitable combination of hand and machine sorting is used. Today, it is possible with the very accurate manoeuvring of machines to perform efficient sorting. However, hand sorting is necessary to finish and control it. Bulky materials such as concrete and masonry are piled on-site and removed to recycling destination (see Table 3.4).

3.3.4 Finishing

Finishing including clearance, levelling the site, removal of all materials and worksite facilities will complete the selective demolition work. The site has to be cleaned and levelled with respect to the agreed end state and the future use of the site. In some cases, the site should be converted to a green area with grass and planting. Recycled materials from the demolition might be used for backfill and levelling of the site.

3.4 CDW MANAGEMENT

After demolition and sorting, the materials from the demolition are handled according to national legislation and local requirements. In general, the CDW management comprises:

- Reuse of bricks, stones, doors, windows, plates, iron bars, timber, etc.
- Recycling of stones, concrete and masonry, recovery of scrap metal
- Recovery of scrap metal
- Disposal of earth and mixed waste, which cannot be reused or recycled
- Special treatment of hazardous waste
- Logistic optimization
- Documentation of the handling and transport of the materials

Provided that the stones, asphalt, masonry and concrete are not polluted, these fractions are 100% recyclable. So is iron and steel, and to lesser extent wood, gypsum and mineral wool. On this basis, it is assumed that 95% of all CDW (excluding soil) in most countries is potentially recyclable. Only 5% must either be incinerated or disposed of at a controlled disposal site. Obviously there is much to be gained by selective demolition and sorting of waste at the demolition site. In this section, a brief introduction to the key issues of CDW management follows; a detailed description of reuse and recycling materials is presented in Chapter 4.

3.4.1 Reuse

A major part of the demolition materials can be reused. Reuse of bricks is very common all over the world, assuming that the bricks are carefully handled during the demolition processe. In addition, doors, windows and plates can be reused, depending on their quality. Old timber of good quality is highly valuable, especially in regions where there is a lack in wood. In less developed countries, it is common to retrieve reinforcement from concrete structures to reuse the bars for new constructions.

3.4.2 Recycling and recovery

Crushed stones, concrete and masonry can be crushed and recycled for materials in new structures, typically road materials. During the last decades, much research and development has been carried out on recycling concrete as aggregates in new concrete. The EU and many other countries have specifications for the use of recycled concrete and masonry.

Recovery of metal is an economically important part of the demolition process. The market price of scrap metal follows the world market for metals. In case of high amounts of scrap metal, the demolition contractor might see his interest in taking out and selling the scrap metal from a demolition project as early as possible, which might affect the demolition process inappropriately.

3.4.3 Disposal

All materials which are not reused, recycled or recovered have to be disposed or treated. The part of the total amount of CDW disposed off depends on the tradition and level of demolition development in the concerned country/region.

3.4.4 Special treatment

Environmentally critical and hazardous waste such as oil spills, chemicals, batteries, and so on, and contaminated waste from cleansing work must be treated or disposed of on controlled landfills. All workers must be instructed in safety and health problems related to the handling of hazardous waste. It is important that they use personal protection gear.

3.4.5 Logistic optimization

After selective demolition and sorting on-site, the materials are transported according to their destination. Some materials for reuse might be sold or picked up by private costumers on-site, while other materials are transported by haulage companies to industrial sites or specialized recycling companies for treatment and transformation into secondary raw materials. The remaining part of the waste is transported to the nearest and cheapest disposal site.

In the EU, the United States and other industrialized countries, the transport cost and tipping fees are a major part of the total demolition costs. In major cities, the transport take many hours and the tipping fees are high, depending on the distance to the city centres. Therefore, the demolition contractors and haulage companies always look for the cheapest CDW handling and transport solution. Illegal solutions such as fly-tipping at random places take place occasionally. Therefore, detailed planning of CDW handling and logistic optimization is needed for reduction of the CDW handling and transport cost. Examples of logistic optimization are presented in Chapter 5.

3.4.6 Documentation

Documentation of the CDW stream is important with respect to correct handling of the waste, as well as documentation of the quality and purity of the materials. According to EOW criteria and qualification of the materials for reuse and recycling, traceability of the waste origin and handling processes are required. It is expected that the traceability will be generally required by the European Commission.

3.5 CLEANSING TECHNOLOGIES

Before the start of selective demolition, the building must be cleansed for all kind of contaminations comprising organic compounds, metal compounds, asbestos and other substances, which are harmful to the environment, health and safety.

The cleansing methods according to the IRMA project (IRMA 2006), listed in Table 3.5, depend on the structure, the surface of the structure and the type of pollution.

Table 3.5 List of cleansing methods and their characteristics

Type of tool	Tools
Mechanical	Milling Abrasive grinding Particle blasting (dry) Particle blasting (moist) Shot blasting Needle descaling
Mechanical hydraulic	High-pressure water jetting Steam jet blasting
Thermal and others	Flame descaling Liquid nitrogen cooling Dry ice blasting Cleansing with chemical agents

Source: Integrated decontamination and rehabilitation of buildings, structures and materials in urban renewal (IRMA 2006).

3.5.1 Milling

The milling technique is predominantly applied for large even surfaces. There are, however, a wide variety of different tools also for smaller areas and for indoor applications. Tools mostly used are steel lamellas, ribs, plates or disks on rotating shafts, or parallel arrays of diamond-tipped saw blades. The advantages of the technique lie in the possibility of the precise adjustment of abrasion depths and widths, and the applicability on almost all types of surface materials of different hardness. In the cleansing context, the milling technique can be used for the removal of different types of coatings from paints to resins and bituminous materials. One of the main disadvantages is the intensive formation of dust. These emissions have to be controlled by suction devices, which can lead to costly constructions. The dust production can also generate the necessity for subsequent cleaning of machined surfaces. Other disadvantages are the transmission of vibrations to sub-surface materials, which can cause material degradation. Furthermore, these techniques are not applicable to uneven surfaces and narrow sites.

3.5.2 Abrasive grinding

The grinding technique is quite similar to the milling method. Fields of application are the smoothening and cleaning of floor and wall areas. Tools are grinding cylinders with fused corundum, grinding discs with embedded diamonds or carbide plates. The ablation depth is lower compared to the milling techniques. In return, the grinding tools are more flexible in use in terms of sites that are difficult to access. Aside from the possibility of depth dependent abrasion, the grinding tools exhibit also the potential to remove hard materials. Due to a wide variety of special grinding tools, the grinding technique is very versatile regarding the processed materials. Disadvantages are that uneven surfaces cannot be treated, materials can be damaged through vibrations and large volumes of dust are produced. Special suction devices can control the latter.

3.5.3 Particle blasting (dry)

The main applications of dry sand blasting are the ablation of concrete surfaces (up to 3 mm), the removal of surface pollutions, coatings, paints, adhesives, resinous and mineral plasters, treatment of natural stone and the blasting of metal surfaces. Fine-grained hard materials are used as blasting agents such as sand, slag, steel grit, corundum or glass beads. Usual grain sizes are 0.5–1.5 mm. The particles are accelerated by pressurized air and hit the surface at angles of about 30–60 degrees. Sand blasting without an enclosure is applicable also for uneven surfaces and narrow sites. In contrast to milling and grinding techniques, the particle blasting does not transfer shocks or vibrations to sub-surface materials. The open application of blasting technique is, however, not recommendable in terms of environmental aspects, as it produces large volumes of contaminated material and the generated dust are spread widely. The dust problem can be confined by a housing that isolates the treated areas and working site from the environment; the waste problem nevertheless persists.

An alternative is the application of the *Vacublast technique.* Here, the produced dust is collected directly at the point of formation by a vacuum device. A second advantage is the recycling of the blasting material in a cyclone separator, which leads to a far lesser production of waste materials. Disadvantages are the higher costs of the method and a minor blasting performance. The Vacublast method is mainly applied for smaller areas. A general problem of particle blasting techniques is the control of the abrasion depth.

Sponge Jet abrasive blasting is an abrasive dry blasting technology based on recycled blasting media. Test data comparing conventional and Sponge Media abrasives have shown that Sponge Media abrasive blasting suppresses up to 99.9% of what normally would become airborne dust. Sponge Media abrasives are manufactured with tough porous urethane sponge material, which controls or suppresses dust. Sponge Media particles flatten as they strike the surface, then expose the abrasive where they cut into the coating and substrate, profiling 0–100 + microns (0–4 + mils) if needed. As the Sponge Media abrasives rebound, the porous urethane creates suction, entrapping dust.

3.5.4 Particle blasting (moist)

The additional use of water in the particle blasting technique leads to a dust reduction of up to 98%. Consequently, the dust suction devices can be much smaller. The water is added directly to the blasting agent in the reservoir, or through an injector in a mixing chamber positioned upstream of the blasting nozzle. With suitable equipment it is possible to combine the dry and moist application of particle blasting. The combination of the techniques can eliminate the problem where particle blasting with additional water can lead to a partlial adhesion of dust and blasting agents on the treated material.

3.5.5 Shot blasting

The shot blasting technique works with steel balls, which are flung at the surface by means of a fast rotating wheel. The steel balls are recycled by a magnetic separator. About 90% of the produced dust is collected by the recirculation. The shot blasting is not suitable for surfaces formed by low-strength materials like plasters. For such materials the degree of abrasion cannot be controlled sufficiently. The method is predominantly applied to the surface removal from floors (concrete, asphalt), or natural stones. The removal of adhesives from screeds strongly depends on the physical characteristics of the adhesive.

3.5.6 Needle descaling

The needle descaling technique is suitable for the removal of weak concrete layers, coating residuals and surface soiling. The advantages of the method are a low formation of dust, a low noise level, a low production of waste and no production of contaminated runoff. However, needle descaling is only applicable for small areas.

3.5.7 High-pressure water jetting

The main applications for this technique are the ablation of concrete surfaces up to the depth of the reinforcement, removal of fire damages, cleansing of natural stonework and other mineral surfaces (e.g. due to oil pollution), removal of resinous and bituminous matter, paints and plasters. The high-pressure water blasting method can be adjusted to the particular purpose by the variation of the water pressure, water temperature (also: steam jet blasting), nozzle characteristics (rotating nozzle beams, point blaster, etc.) and additives (sand, detergents, etc.). There exist applications with or without reuse of the water. As the water demand of the method is extremely high, from an environmental point of view the non-recycling option is not recommendable. Moreover, it bears the risk that the contaminated wastewater infiltrates into other parts of the building. On the other hand, the use of a suction apparatus with a housing, which is sealed to the processed area, limits the performance of the system considerably.

Compared to particle blasting techniques the water blasting techniques are more expensive. They have a high water demand that leads to large amounts of wastewater with contaminants in high dilution. On the other hand, there is no production of dust.

3.5.8 Flame descaling

In the context of cleaning, flame descaling can be used for the removal of old paint coatings (other application: removal of concrete surface layers). Whereas the removal of concrete can be regarded as a non-polluting method, the flame scaling of coatings can induce a variety of highly problematic degradation products of organic components. Moreover, the treated surface has to be reprocessed by further cleansing methods after the application of flame scaling. From the viewpoint of safety at work, the technique implies a fire hazard and the risk of intoxications through produced gases.

3.5.9 Liquid nitrogen cooling

The application of liquid nitrogen is suitable for the removal of coatings, PCB-containing joint sealants and plasters. The main advantages of this method are the absence of dust and noise emissions, vibrations, the absence of residues of the treatment agent and the avoidance of damages to the sub-surface layers. The main disadvantages are the high costs and the confined field of application, as the method cannot be applied to homogeneous materials (removal of concrete surfaces, etc.).

3.5.10 Dry ice blasting (non-abrasive)

The blasting agent is solid CO_2, which is pressed to pellets and blasted on the surface by a special gun. The cleansing effect is based on the thermal cracking of the surface material (coatings, dirt on surface, dyes) and the partial removal of cracked matter by the expansion of CO_2 through evaporation. The dry ice blasting technique is applicable to craggy surfaces and fit for the removal of soft/flexible materials. It holds no risk of damages to sub-surface materials through vibrations or concussions. From the waste management point of view, it has the advantage that no toxic residuals of treatment agents are produced. Like the treatment with liquid nitrogen, it is comparatively expensive and not applicable to homogeneous materials.

3.5.11 Cleaning with chemical agents

The cleaning of surfaces with agents is mostly applied to the removal of paints and coatings, and typical in the field of graffiti treatment. Water-based treatments with added agents, like detergents or complexing agents, can be carried out by handcraft or combined with water blasting techniques. Yet the necessity to collect and recycle or dispose of the runoff is even higher than with water blasting techniques. The work with solvents holds the additional necessity of extensive safety and health measures. As many solvents are volatile, partly or completely water-soluble and at the same time noxious or toxic, they also carry a high environmental risk.

3.5.12 Development of cleansing techniques

The development of cleansing techniques concentrates on two main targets: the improvement of the separation performance for contaminated and uncontaminated materials, and the reduction of emissions – either by suitable retention devices or by modification of the removal technique itself. The particle blasting methods have to be improved, particularly in terms of the consumption of blasting agents. The possibility of recirculation of agents strongly depends on the physical characteristics of the used materials (hardness, density). An important item concerning emissions is the diffuse particle emission during application.

The water blasting techniques have to be optimized with regard to the problem of water intruding into adjacent constructions. Linked to this demand is the need to reduce the water consumption. Housing of the treated areas and recycling of water and debris will be investigated with the aim of further enhancement. The thermal and chemical techniques are predominantly problematic due to gaseous and/or liquid emissions of solvents, or the release of decomposed organic matter. These emissions have to be monitored, and modifications of the techniques or the retention devices have to be tested.

3.5.13 Decontamination of waste from cleansing

The decontamination of wastes from cleansing procedures performed on contaminated structures applies to solid and liquid wastes mainly. The decontamination of the solid fraction can be handled in a manner analogous to the decontamination of mineral deconstruction wastes. For the liquid wastes, methods of wastewater treatment are to be considered.

The decontamination of deconstruction waste often is performed with the same facilities as for the cleaning of contaminated soils. In many cases, deconstruction wastes and contaminated soils are produced at the same time. The typical contaminants also correspond. Customary methods for the cleansing of soils and mineral wastes are washing procedures with water and additives, thermal methods (pyrolysis), bioremediation, and soil-air suction and cleansing of the circulated air. These methods mostly apply to the removal or decomposition of organic pollutants. For the removal of heavy metals the pyrolysis, bioremediation or air suction are not expedient. Furthermore, the thermal and bioremediation procedures are only ill-suited for on-site deconstruction measures. The experiments consequently will focus on washing procedures with a variation of agents. These procedures lead directly to the decontamination of wastewaters from cleansing measures, which also have to be considered separately for inorganic and organic contaminants. Heavy metals can be removed by different procedures such as precipitation, ion exchange or adsorption on special agents. Organic pollutants can be biologically degraded by microorganisms. For on-site measures, however, an oxidative chemical procedure is easier to perform (e.g. oxidation with peroxide with simultaneous UV irradiation) and also is fit to decompose persistent pollutants like organo-chlorine compounds.

3.5.14 Asbestos removal

Asbestos and building materials containing asbestos have been used in the building and construction industry from the end of the 1920s to the end of 1970s. Asbestos is found in roofing plates, covering plates, insulation plates, fire protection products, and so on. Asbestos dust is toxic and has a serious impact on human respiratory passages and lungs.

Dusty asbestos contamination is cleansed using the above-mentioned techniques in an airtight work chamber. Hard, non-dusty asbestos materials, such as roof plates are removed by careful dismantling without breaking or damaging the plates. In cases where the plates break, the removal should take place like the removal of dusty asbestos (Figure 3.3).

Figure 3.3 Illustrations of decontamination methods and tools. Left: Remote surface cleansing of a 150 m high smokestack before demolition (see Figure 3.6). The paint contained PCB. (Photo courtesy of Jonny Christensen, Municipality of Copenhagen). Right: Surface cleansing of a contaminated wall. (Photo courtesy of NIRAS DEMEX.)

3.6 DEMOLITION OF CONCRETE

Disintegrating and crushing concrete structures is one of the major challenges of demolition operations. The demolition contractor has a toolbox of applicable means for the decomposition of concrete comprising of:

- Crane and ball
- Breaking/hammering
- Crushing
- Wedding/bursting
- Blasting/explosive demolition
- Abrasive drilling and cutting
- Other technologies
- Developing technologies

The most common tools, see Table 3.6, are presented briefly in this section with reference to the IRMA project (IRMA 2006). Further details are found in the *Demolition and Recycling International Buyers' Guide* (D&RI 2017) and the individual producer's information.

3.6.1 Breaking mechanisms

Concrete consists of stones, sand, water, air and cement plus some additives. It is a characteristic of concrete that the compressive strength, typically 20 Mpa–50 Mpa, is tenfold higher than the tensile strength. In order to compensate the weaker tensile strength, iron bars are embedded to take up

Table 3.6 Brief characteristics of the most common demolition techniques

Demolition technology	Primary application	Environmental impact	Comments
Crane and ball	Total demolition of buildings and structures, up to heights of 50 m	Dust, noise and vibrations	Requires ample space
Concrete crushers/breakers	Total and partial demolition of reinforced concrete structures, plates, columns and beams	Dust, noise and vibrations to be controlled	Many sizes and types of tools. Can be mounted on a long-reach machine, reaching 50 m
Handheld hammer	Small-scale demolition and removal of concrete, repair of concrete structures, shortening of foundation piles, etc.	Dust, noise and vibrations Heavy personal workload. Risk of white finger disease	Daily work time limited to shorter periods according to EU directive (EU 2002)
Hydraulic hammer	Total and partial demolition of all kinds of concrete structures	Dust, noise and vibrations	Many sizes and types of tools. Can be mounted on a long-reach machine, reaching 50 m
Hydraulic bursting	Small-scale demolition of plain concrete structures	Noise, dust and water from drilling	Require preparatory drilling work. Time consuming and expensive
Blasting	Turning over of tall structures. Demolition of foundations and big reinforced concrete structures. Mini Blasting of small reinforced concrete structures	Short time impact of dust, noise and vibration from blasting. Dust and noise from preparatory drilling work	Specialist work Provenance and handling of explosives require authorization
Abrasive drilling and cutting	Hole cutting in reinforced concrete structures, cutting of concrete structures for separation and dismantling of greater structures	Noise, dust or water depending on the cutting type – dry or wet	Well defined work performance and accurate cutting boundaries
Hydrojet	Surface treatment of concrete, exposure of reinforcement bars	Water spill, risk of accidental impact of the high-pressure jet	Accurate work performance Specific safety procedures are needed

the tensile strength. Demolition of concrete requires the impact of forces exceeding the strength of the concrete and the introduction of fracture energy enabling the decomposition of the concrete. In case of overloading a reinforced concrete structure, the iron bars take the tensile forces and a ductile breakage occur. In the case of exceeding the tensile strength of plain concrete, a separation breakage occurs and the concrete structure decomposes.

The fracture mechanism, illustrated in Figure 3.4, depends on a number of concrete parameters: compression strength, tensile strength, cement concrete, quality of aggregates, porosity, compacting, and so on. The applied concrete demolition technology should be based on impact forces exceeding the fracture energy of concrete with respect to compression strength or tensile strength. Direct impact on the concrete using crushers, balls, hammers, high-pressure water jet (hydrojet), and so forth, requires an impact strength exceeding the compression strength. Demolition of high-strength concrete, with a compression strength of more than 50 Mpa, using hammers or crushers/breakers might be very hard on the tool and

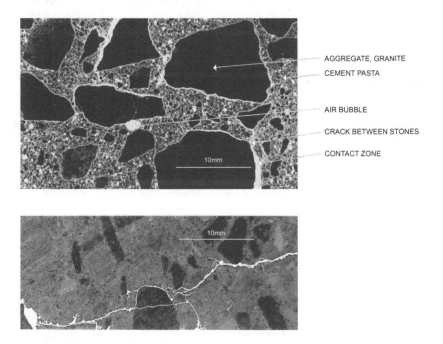

Figure 3.4 Thin sections of concrete show fracturing of concrete. The fractures start in the weaker layer between the cement particles and continue fracturing into micro-cracks, macro-cracks and disintegration of the structure. The picture above shows micro-cracks in the cement pasta between the stones. The picture below shows cracks between the stones, around the stones and through the stones. (The thin sections and pictures were prepared by PELCON, Denmark.)

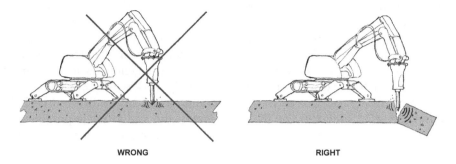

WRONG   RIGHT

Figure 3.5 Breaking of concrete. Demolition of a heavy, high-strength concrete structure must start from a free edge of the structure, where the introduced crack energy can reach the surface and cause the disintegration of the structure.

time consuming. Abrasive drilling and cutting in concrete with aggregates of flint causes the need for frequent changes of the abrasive tools, which makes the technology expensive and time consuming. Indirect impact using chisels/wedges, expanding tools, blasting, and so on, requires impact strength exceeding the tensile strength. These methods might be preferred to demolish high-strength concrete compared with the surface impact tools. See Figure 3.5.

The economy and total energy consumption, including CO_2 emission, for demolition of a concrete structure depends on the required sizes of the broken concrete materials (rubble or debris). In general, demolition of concrete into small size debris is more expensive and energy consuming than demolition into bigger sizes. This is also true when speaking about crushing concrete debris for recycling.

3.6.2 Crane and ball

Crane and ball is an older, traditional demolition technique using a crane with a swinging or falling ball, weighing from 1 t to 7 t. The technique is used for higher demolition objects, normally up to 30 m in height and with extended jibs, up to 50 m. The technique is effective, but very noisy and dusty. Furthermore, the crane operation is very space consuming. Today in industrial countries, the crane and ball has been replaced by concrete crushers mounted on long arms. Therefore, the use of crane and ball is very restricted and limited.

3.6.3 Breaking/hammering

Demolition by chopping/hammering is the most common method for concrete demolition. Like crushing/braking/cutting, various types and sizes of hammers (jackhammers or breakers) are found on the market. Small hand-operated pneumatic, hydraulic and electric hammers, weighing

5–50 kg are commonly used for demolition and removal of smaller amounts of concrete, such as spills of concrete, removal of concrete for the repair of concrete structures, providing holes in structures, cutting concrete slabs, shorting pile tops, and so on. It should be noted that handheld hammering can be critical to workers' health, because of the general workload and risk of white finger disease. Therefore, the daily work with heavy handheld hammers should be limited to short periods, as recommended by the EU Directive of 25 June 2002 on the minimum health and safety requirements regarding the exposure of workers to the risks arising from physical agents (vibration) (EU 2002).

For the demolition of concrete structures, hydraulic hammers mounted on excavators are used. All sizes of excavators and hammers are found from mini-excavators, 500 kg–5 t, to large-scale wheeled and tracked excavators up to 100 t and heavier mounted with breakers from 20 kg to more than 10 t. Breakers can be mounted on long-reach excavators with an operating range/height of 50 m (see Figure 3.6).

3.6.4 Crushing

For crushing concrete, various types and sizes of crushers are found on the market. The mechanism is based on pressure impact provided by a hydraulic operating steel jaw exceeding the concrete compressive pressure.

Small hand-operated crushers, with a weight of 25–100 kg, can crush concrete walls and structures of 100–350 mm in thickness into small bits. Crushers, weighing 500 kg–10 t, mounted on excavators can crush heavy concretes structures with a thickness of up to 1.5–2 m. For demolition of high structures up to 50 m in height, crushers mounted on long-reach machines can be used (see Figure 3.6). The operators control the machines using a video system with a camera mounted on the top of the machine.

Today, the concrete crusher is considered as the preferred demolition machine compared to the hydraulic breaker/hammer and crane and ball because of its reduced environmental impact from noise, dust and vibration.

Multi-processors are attachments with a combined crushing and cutting function for demolition purposes.

Pulverisers are attachments like crushers primarily for downsizing big concrete debris after demolition into small debris for recycling.

3.6.5 Wedging/bursting

In the past, breaking of rocks and stones was performed by the use of iron wedges in drill holes providing tensile force inside the rock resulting in fractures and disintegration of the mass. Similar technologies are used for concrete demolition. After drilling boreholes of 100–200 mm diameter, a hydraulic or pneumatic wedge tool is placed in the hole and causes cracks by expansion of the wedge. After drilling a number of holes side by side,

Figure 3.6 Examples of demolition of high-rise structures. (a) Shows the demolition of a football arena by a crusher mounted on long-range machine. (b) Shows a 150-high chimney and (d) shows the demolition of the chimney by use of a remote-controlled demolition hammer mounted on a movable platform. The text on the platform (c), says 'Goodbye. I will be recycled in the Sydhavnen new recycling centre'. Half of the 5,000 t concrete will be recycled as aggregate in new concrete structures. (Photos Jonny Christensen, Municipality of Copenhagen, and Danton A/S.)

parts of the concrete structure are detached successively. This technology is very controllable but time consuming and expensive because of the need for diamond drilling.

Expansive chemical agents and expanding mortars are available on the market. The mortar, for example composed by 80% lime, 10% siliceous sand and 10% retarding agent diluted in water is poured into drill holes, typically 32 mm in diameter; after hardening, the mortar expands and produces cracks in the concrete. The drill holes are made by hydraulic or pneumatic drill machines. A short distance between the drill holes is required. The drilling and the expansive mortars are quite expensive and the hardening takes time (from 30 minutes to hours).

The splitting/bursting technologies are time consuming and expensive. Also, they are not appropriate for demolition of reinforced concrete. Therefore, the splitting/bursting technologies are not competitive compared to other concrete demolition technologies.

3.6.6 Blasting/explosive demolition

In cases of very thick constructed structures or high structures, demolition using explosives placed in drill holes can be applied. Demolition of buildings by blasting can be performed as overturning of the structure or as vertical progressive collapse called *implosion*; see examples in Section 3.7 and Figures 3.11 and 3.12.

Explosives are substances capable of releasing their potential energy within a short period of time through a violent chemical decomposition triggered by the action of heat or an impact. For demolition of small concrete and special concrete structures, for instance in cases when repairing existing concrete structures, *Mini Blasting* (Lauritzen and Petersen 1991; Molin and Lauritzen 1988) is very useful. Mini Blasting is carried out by the use of very small explosive charges, 5–50 grams, and complete safety covering, which enables blasting on normal worksites with a 5–10 m safety distance to other workers (see Figure 3.7).

Figure 3.7 Examples of Mini Blasting. (a) Shows cutting of window holes by blasting. (b) Shows concrete replacement in a bridge column keeping the old undamaged reinforcement bars to be reused. (From Lauritzen, E. K. and Petersen, M., Partial demolition by Mini-Blasting. *Concrete International,* June 1991.)

Blasting is performed by specialists who are trained and licensed according to national rules. The provenance and use of explosives are subjected to specific EU and national rules on the control of explosives to minimize the risk of terror actions.

3.6.7 Abrasive drilling and cutting

This procedure involves the cutting with blades, wire and drilling with tools provided with industrial diamonds-tipped tools or carborundum (silicone carbide, SiC) tools. Usually a lot of water is needed for cooling and spraying the cutting area. Dry cutting techniques are also available. An evacuation for the used water is required. The system to remove the pieces is an important aspect of this procedure, and of course the size of such pieces, which greatly influence the final cost of the operation. Diamond cutting is a very precise demolition method and leaves a defined edge on the structure. Compared to other technologies, diamond cutting is rather expensive and time demanding.

3.6.8 Other technologies

Different systems are used in particular situations, as a complement to ordinary mechanical and hand demolition procedures. Most of them are methods presently in the process of being developed and investigated for improvement and implementation in the future. Among these are methods like expansive mortar, water jets, lasers and others.

Thermal cutting and drilling using oxygen torches is based on an exothermic process, which brings the concrete and steel to a melting point at 2,000–2,500 degrees Celsius. The oxygen torch is improved by using a fine particle powder of a mixture of iron (60%–85%) and aluminium (40%–15%). These particles burn on the periphery of the oxygen cutting jet, increasing its heat.

High-pressure water jet blasting (hydrojet) and cutting is used in the demolition industry for cleansing reinforced concrete surfaces and removal of damaged concrete before repairing the structures. The industrial use of water jet blasting is based on a pressure of the water jet of 1,400–4,000 bars and a nozzle diameter of 0.12–0.5 mm.

3.6.9 Developing technologies

In recognition of the need for improved concrete technologies, development in this area has taken place all over the world. This development aims at the disintegration of reinforced concrete, as well as remote controlled robot technologies, especially for the decommissioning of nuclear power reactors in the United States, Japan and Europe. During the planning and implementation of the *Japan Power Demonstration Reactor* (JPDR) decommissioning program 1986–1996, various cutting techniques for dismantling the reactor pressure vessel and the biological shield were tested (Yokota et al. 1993). Furthermore, other new technologies have also been were also tested:

- Laser
- Electric heating of iron bars in reinforced concrete
- Electric induction
- Microwaves

Experiments with CO_2-lasers have resulted in the demolition of concrete to a depth of 200 mm from the surface. Applying a voltage of 400 V and an amperage of 250 A, reinforcement bars can be heated up to 400°C–450°C, which caused fractures and spalling of concrete (see Figure 3.8). Electric heating of a reinforced concrete object can also be applied by the use of an electromagnetic device mounted on the concrete surface, enabling an induction current in the reinforcement bars. Experiments using microwaves (see Figure 3.8) have resulted in the destruction of the concrete surface to a depth of 150–200 mm.

These technologies are advantageous compared to traditional technologies with respect to noise, vibration and dust. However, they are very energy consuming and the hardware is expensive. It should also be mentioned that through the ages a range of concrete demolition technologies based on chemical reactions, temperature cycling, infrared radiation, ultrasound impact, and so on, have been invented and tested. However, much research and development is needed before these technologies can be put on the market.

Figure 3.8 Left: Test of demolition by electrical heating of the reinforcement bars, resulting in cracks and disintegration of the concrete. Middle: Picture shows the cracks around the heated reinforcement bars. Right: Demolition of concrete by microwave heating.

Figure 3.9 Logo of the UK Institute of Demolition Engineers. (With permission of the Institute of Demolition Engineers.)

Presentations about research and development of demolition technologies are found in the proceedings of the second and third international symposium held by RILEM (the International Union of Testing and Research Laboratories for Materials and Structures) on demolition and reuse of concrete (Kasai 1988; Lauritzen 1993).

The walls of Jericho were not demolished by the use of ultrasonic horns, and no evidence of a sudden destruction of the walls has been found. Referring to Chapter 6, looking at the effects of natural and technical disasters, we must realize that the technical capabilities of demolition are far from the natural capabilities of destruction. An earthquake might have caused the Jericho demolition. However, the Jericho event might inspire demolition engineers into creative thinking for alternative solutions. This is the case for the members of the UK Institute of Demolition Engineers (see Figure 3.9).

3.6.10 Health and safety

Demolition workers have in general a dangerous working place because they move about in an area where big machinery is used and heavy materials are moved about. In order to avoid potential risks, it is of the greatest importance

- That the demolition worker and the machine driver are in close communication and can rely on each other

Table 3.7 Identification of workers' health and safety risks depending on demolition technique

Workers, health and safety risks	Demolition technique			
	Hand demolition	*Demolition by machines*	*Demolition by explosives*	*Cutting, drilling and sawing*
Dangerous materials	x	x	x	x
Dust	x	x	x	x
Crash	x	x	x	x
Electrical shock	x	x		x
Lack of security railing	x	x		x
Lack of pers. protection equipment	x	x		x
Vibration	x	x	x	x
Use of tools	x	x		x
Exhaust fumes	x	x		x
Heavy lift	x			
Noise	x	x	x	x
Thermal impact	x	x	x	x
Weak positions	x	x		x

Source: Integrated decontamination and rehabilitation of buildings, structures and materials in urban renewal (IRMA 2006).

• That most of the total demolition work should be conducted from outside of the workplace, and dangerous workplaces on the structure should be avoided
• That workplaces on the structures must be safeguarded

It is the responsibility of the machine driver to size up the static stability of the building under demolition and to ensure that no one is hurt either by the machine or by debris. On the other hand, the demolition worker has to rely on and also obey the directions given by the machine driver. An unproblematic communication is of great importance.

Typical health and safety risks related to specific demolition techniques/ work methods are presented in Table 3.7, comprising hand demolition, demolition by machines, demolition by explosives and abrasive demolition technologies.

3.7 DEMOLITION METHODOLOGY

The design of a demolition, dismantling or decommissioning project depends on a number of various factors: type of structure, scope of the job, location, surroundings, environmental conditions, urgencies and political matters. An exhaustive analysis of all these circumstances will be fundamental, when facing a process of this type, to secure the major guarantees of success. Furthermore, with reference to the objective of this book, it is important that the demolition design does not only aim for the removal of the structure, as the design must also take into consideration all aspects for the most optimal handling and quality of recycling of CDW.

In this section, the most typical demolition jobs will be discussed followed by examples of actual interest, including:

- Buildings
- Towers, chimneys, silos, etc.
- Bridges

3.7.1 Demolition of buildings

Total demolition of buildings should be carried out as selective demolition, as presented in Section 3.3. After cleansing and stripping, the demolition of the building structure takes place as follows:

- *Buildings with a height below approximately 10 metre* (3 storages) are demolished by means of an excavator with bucket. In case of hard concrete structures, the bucket might be changed with a concrete breaker
- *Buildings with a height between 10 and 30 metres* are demolished by crane and ball, or long-reach excavators with buckets and/or breakers and hammers. Alternatively, the demolition takes place from the top with machines working on platforms made of demolition rubble
- Buildings with a height above 30 m are
  - Demolished from the top with small machines
  - Demolished by long-reach machines, normally up to a height of 50 m
  - Demolished by crane and ball, normally up to a height of 50 m
  - Demolished by blasting
  - Disassembling from the top by cutting or separation of elements, which are taken down by crane

Demolition from the top with small machines is applicable for buildings with very narrow space around them. Light machines, excavators with breakers and/or hammers are lifted to the top and placed on the structure by cranes. This requires that the bearing strength of a structure and the decks acting as work platforms have been controlled. Dynamic forces from the workload must be included in the bearing capacity. Successively, the structure is demolished from the top and the CDW is taken to containers on the ground in dust-protected waste pipes, inside or outside the building. Often the building will be coved completely, which keeps all dust inside and contributes to noise dampening.

The advantage of this method is that the procedure can be controlled. Presupposing an adequate and careful control of the worksite, it has a high degree of safety and a low environmental impact on the neighbours. The disadvantages are the time and cost of the demolition process.

Demolition by long-reach machine attached with breakers or crushers is applicable for buildings up to approximately 50 m high. Operation of the long-reach machine requires sufficient space around the building. It is difficult to cover the site, which requires special means to reduce the dust and noise.

The advantage of this method is that it is effective and economic. The disadvantage is that the work height and the shaking of the structure during the demolition work might cause risks of unintended collapse of the structure and the uncontrolled falling of debris. Control of dust and noise require special attention as well.

Demolition by crane and ball is also applicable for buildings up to approximately 50 m in height. Operation of the crane requires sufficient space around the building. Like the long-reach machine, it is difficult to cover the site, which require special means to reduce the dust and noise.

The advantage of this method is that it is very effective and economic. The disadvantage is that the work is very dusty and noisy. The demolition work might cause risks of unintended collapse of the structure and the uncontrolled falling of debris. In most EU countries, the crane and ball is deselected in favour of the long-reach machine because of environmental concerns.

Demolition by blasting is commonly used for the demolition of high-rise structures above 50 metres. The blasting design aims on vertical collapse (implosion), turn over of the building or a combination of turn over and implosion. Normally, the blasting is carried out with explosives placed in the bearing concrete structures of the ground floor and of a few selected stories to enable the predicted collapse mechanism.

The advantages of blasting are the very short time of the physical demolition of the structure and removal demolition, and the resulting environmental impact. The disadvantages are the short but very intense impacts of dust, noise and ground vibrations. The preparations for the blasting including drilling, placing of explosives and the comprehensive safety management take much time. Finally, blasting is associated with the risk of uncontrolled collapse, and damages to neighbours from debris and dust. After all, blasting demolition is a very competitive demolition method for tall buildings and structures with respect to price as well as environmental protection.

3.7.1.1 Disassembling of buildings

Sometimes, when there is a lack of space, a very sensitive environment, unstable buildings or buildings designed for disassembling or dismantling, the demolition – total or partial – is performed to separate building structures from top to ground in the reverse order of the original construction, as shown in Figure 3.10. After cleansing and stripping the building, including removal of roof and windows, the structure is taken down by crane piece by piece. In accordance to normal building practice, all structural elements are connected with joints of reinforcement bars in concrete or mortar, which are not designed for disassembly. Opening the joints or separating the elements requires cutting techniques (abrasive cutting, Mini Blasting or chopping/crushing by small machines on the structure).

Compared to traditional demolition methods, presented in earlier sections, the dismantling method is more expensive and takes more time. On the other hand, the method is much more environmentally appropriate, and the method is applicable when aiming to reuse building elements. The dismantling method will be mandatory for demolition of buildings specially designed for disassembly, which is a common EU goal for future constructions (Table 3.8).

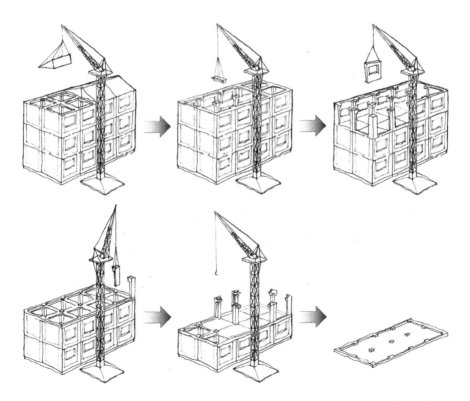

Figure 3.10 Principle of disassembling of high-rise structures by crane lifting. (From Lauritzen, E. K. and Jacobsen, J. B., Nedrivning og Genanvendelse af bygninger og anlægskonstruktioner, SBI Anvisning nr. 171, 1991 [In Danish].)

Table 3.8 General comparison of the qualities of methods for demolition of high-rise structures with respect to risk and safety, environmental impact, work time and economy

Methodology	Risk	Environment	Time	Economy
Demolition from the top	xxx	xxx	x	x
Long-reach machines	xx	xx	xx	xx
Crane and ball	x	x	xx	xx
Blasting	x	x	xxx	xxx
Disassembling	xxx	xxx	x	x

Note: xxx: most preferred, x: less preferred.

3.7.2 Risk-based demolition design

The selection of the design of demolition for high-rise buildings and structures should be based on a combined economic, environmental and risk assessment approach. For the assessment of the possible demolition methods a risk-based approach is recommended using the semi-quantitative analysis method. The risk is assessed as the combined result of

- Likelihood of hazardous occurrence, see Table 3.9
- Severity of consequence, see Table 3.10
- Level of risk, see Table 3.11

The methods have been evaluated by initial and residual risk according to the ALARP principle (as low as reasonably practicable). For a risk to be ALARP, it should be possible to demonstrate that the cost involved in reducing the risk further would be grossly disproportionate to the benefit gained. The likelihood of hazard occurrence for a given critical event was categorized into five levels (see Table 3.9). The likelihood of hazard occurrence for a given critical event is categorized into five levels, numbered 1–5 (see Table 3.9). The likelihood of hazard occurrence is measured against the severity of the consequence of an accident on a scale of five levels, marked A–E (see Table 3.10). Based on the combined assessment of the hazard level and the consequence level of a given event, the level of risk can be classified in the risk matrix (see Table 3.11).

Table 3.9 Likelihood of hazard occurrence

Scale	Level
I	Improbable
2	Occasional
3	Probable
4	Frequent
5	Very frequent

Table 3.10 Severity of consequence, example of scale and level

Scale	Level	Occupational safety	Safety, third-party persons	Safety, third-party property
A	Negligible	Short discomfort, no need for treatment	None	None, insignificant damage, no repair
B	Marginal	0–1 sick days	Short discomfort, no need for treatment	Cosmetic damage, easy to repair, no consequences for the function of the building
C	Serious	>1 sick days	0–1 Sick days	Moderate damage, repair possible, no or limited consequences for the function of the building
D	Critical	Permanent injuries	>1 Sick days	Comprehensive damages, repair possible, temporary consequences for the function of building
E	Catastrophic	Death	Permanent injuries, death	Comprehensive damages, impossible to repair or/ and consequences for the surroundings

The risk matrix includes risk levels from 1A to 5E, where the shade of grey, from dark to light, indicates whether the level of risk is 'not acceptable', 'undesirable', 'tolerable' or 'acceptable'. Based on a critical review of all possible events and potential hazards, a list of events/hazards can be prepared with recommended measures to reduce the risk level. See the example of risk evaluation of demolition blasting in Table 3.12, described in Section 3.7.2.1.

Table 3.11 Example of A risk matrix showing risk levels from 1A (lowest) to 5E (highest). classified low risk (white), tolerable, acceptable risk, (light grey), medium, undesirable risk (grey) and high, not-acceptable risk (dark grey)

Risk level		Likelihood of hazard occurence				
		Improbable 1	Occasional 2	Probable 3	Frequent 4	Very frequent 5
Severity of consequence	Neglible A	1A	2A	3A	4A	5A
	Marginal B	1B	2B	3B	4B	5B
	Serious C	1C	2C	3C	4C	5C
	Critical D	1D	2D	3D	4D	5D
	Catastrophic E	1E	2E	3E	4E	5E

Table 3.12 Example of risk evaluation and mitigation sheet

Event/ hazard	Risk level	Possible occurrences	Possible causes	Mitigation measures
Unwanted/ unintended collapse	3E	The building falls in an unintended direction	Error in blast design Error in execution of drilling/ charging	Increased control of the contractors' blast design and check of collapse mechanism. Increased control of contractor during drilling and charging. Use of double-firing system
Building only falls partly	4C	Part of the building is still standing after blasting	Error in the drilling/charging/ ignition Construction stronger than expected	Increased control of the contractors blast design and check of collapse mechanism. Increased control of contractor during drilling and charging. Use of double-firing system
Flying debris	2D	Damage due to falling/ flying/ rebounding debris	Inefficient covering of objects or/and securing of third-party property Insufficient safety zone	Covering of the building subject to blasting by the use of heavy and light blast mats. Increased control prior to blasting. Securing of exposed buildings with plywood plates, net and old containers

3.7.2.1 Example of risk-based demolition, Copenhagen 2012

Two concrete buildings in Copenhagen, Denmark, 13 and 15 storeys each, were constructed in 1954–1956. Because of outdated designs and high costs of renovation, the owner, AAB co-operative housing society, decided in 2006 to demolish the buildings and hired the consulting company *NIRAS* for the planning and management of the demolition work. The two buildings were situated among spread out low residential housing, a kindergarten/nursery, an after school centre and main traffic and public transport routes.

After an initial selection the options for demolition came down to three possible methods, namely, long-reach concrete crusher, deconstuction from the top lifting elements by crane and demolition by blasting. The three methods were evaluated with respect to uncontrolled collapse, occupational safety, third-party damages, environmental impact, time and budget. Based on the risk assessment, several possible hazard occurrences were found and linked mitigation measures were established to minimize the risks according to the ALARP principle.

On 12 May 2012, the two tower blocks in Copenhagen were successfully demolished through blasting conducted by the Danish demolition company *Brandis A/S* in cooperation with the Scottish blasting contractor *SAFDEM*. The demolition was the biggest building blasting job in Scandinavia until now. The total amount of explosives provided by *Orica Mining Services A/S Denmark* was 60 kg of explosives, 30 kg per building, placed in drill holes with 1,600 charges, 800 per building (see Figures 3.11 and 3.12).

Figure 3.11 Example of blasting design for two 15-storey buildings in Copenhagen 2012. (Improved model with permission of INGENIØREN 2012.)

By assessing the initial selected methods on safety risk, environment protection, time and cost it was possible to rank and select the best method for the demolition. Furthermore, the process also included an evaluation of hazardous scenarios and a thorough consideration of the possible linked mitigation measures for the hazardous occurrences. Risk assessment and management was clearly the right tool in selection for the best demolition method. At the same time, it enabled the project management to identify and foresee hazardous occurrences and implement the mitigation measures to address and prevent them.

The case of these blasting demolition projects showed the following:

1. In order to convince the client that the risks are acceptable, the blasting contractor must present detailed documentation of the design of blasting and his experience
2. An open discussion of the risk and co-operation of risk mitigation measures and environmental protection is necessary in order to achieve public acceptance
3. Public awareness and information is important
4. Finally, the results of the blowdown clearly demonstrated that the blasting was the most competitive demolition method

Figure 3.12 Blowdown of the first building, 12 May 2012, in accordance to the design presented in Figure 3.11. (Copyright Jorgen Schneider, Orica Denmark.)

3.7.2.2 *Dismantling of the 22-storey 'Ronan Point' dwelling building, London 1986*

The Ronan Point building, constructed in 1967–1968, was hit by a gas explosion only 3 weeks after being put into use. The explosion occurred in a corner apartment on the 18th floor, causing a progressive collapse of the entire southeast corner of the building. The building could have been sawed and reconstructed. However, it was discovered that the joints of the elements in the building were poor and faulty. The fire safety of the building was not satisfactory, either. Furthermore, at that time in the United Kingdom, a general reluctance was seen to living in tall buildings. Because of asbestos in the building it was not suitable to carry out the demolition by blasting. It was decided to take the building down by dismantling. The concrete elements were crushed by a crushing machine placed on the site.

3.7.3 Partial demolition of buildings

The highest quality of recycling building materials is the reuse of the building. In some cases, the building might be used as is. Therefore, in most cases the building must undergo considerable repair work and changes to be reused, which requires a partial demolition to a minor or major degree, for instance:

- Demolition of backyard buildings, staircases
- Demolition of the building except the basic bearing structure, which will be reused
- Opening in façades for windows or doors
- Interior demolition of walls and bearing structures
- Extension of basement for parking

Partial demolition is carried out mainly by the use of small demolition machines and handheld tools with special regard to the stability of the building and requirements for an accepted environmental impact. Depending on the situation and scope of the work, some parts of the building will be used during the demolition and partial repair. This requires a high control of inconveniences and environmental impacts, especially dust, noise and vibration. Noise and vibration from concrete demolition work is very hard to silence and reduce. In the case of partial demolition of concrete in sensitive environments, such as in an office building where people are working in another part of the building complex, it is recommended to insulate the part to be demolished from the other part of the building by cutting all physical connections in order to stop transmission of vibrations through the structural members.

The reconstruction of the World Bank headquarters on H Street in Washington, DC, in 1989–1997, is an example of problems with partial demolition of an office building with working people. The reconstruction work comprised successive demolitions and reconstructions in the building complex from part to part without stopping all of the office work. One of the major problems was dampening the noise and vibration emissions from the demolition work. A lot of unforeseen planning and insulation work was needed during the reconstruction contract.

3.7.4 Demolition of towers, chimneys, silos, and so on

Usually, tall industrial structures, such as towers, chimneys, silos, and so forth, have a shorter lifetime than dwelling houses. The lifetime depends on the actual situation of the works and the function of the structure. In Europe, the introduction of natural gas has made many local heating plants obsolete. The decline of mining and heavy industries in the developed countries, specifically Europe and the United States, have ended the use of many industrial sites, and resulted in the need for transformation and urban rehabilitation. Following the development and centralization of the food industry, many silo plants have been closed and demolished or transformed for other purposes.

The design of demolition of tall structures is based on the following principles:

- *Tall structures with dense surroundings* are demolished by breaking or crushing by small machines from the top, or dismantling section by section after abrasive cutting
- *Structures with a height below 50 m, with good surrounding space* are demolished by crane and ball, or long-reach crushers or hammers
- *Tall structures, with a height above 50 m, with good surrounding space* are demolished by blasting

Demolition by chopping or crushing requires small machines to be lifted by crane to the top and placed on a climbing platform, which can be lowered successively in time with the progress of the demolition work. Section-wise dismantling needs horizontal cuts in the structure. Usually, the cuts are performed by an abrasive saw or wire tool, which also need a mobile work platform to place the cutting tool. The cuts might be provided by handheld chopping or Mini Blasting.

Compared to successively demolishing or dismantling from the top, blasting is much quicker. An example of blasting of a chimney is shown in Figure 3.13. However, turnover of tall structures by blasting requires at least free space in the falling direction of 1.5 times the height of the structure. Blasting can also be performed by shortening the fall or implosion.

Figure 3.13 Very accurate blasting of a 60-m high concrete chimney, Denmark 1989. With only 2 m distance to new installations on the right side and 3 m distance to the new chimney, the blasting was planned and executed with extreme control over the falling direction. (Photo: NIRAS DEMEX.)

Demolition of high-rise structures requires special attention with respect to risk-based demolition design, as seen in previous section. Because of special operations and extreme work conditions, demolition of tall structures needs careful planning and performance-minimizing risk of critical incidents. Examples of typical incidents are:

- Unintended turnover or vertical collapse of structure by blasting
- Loss of load during crane lift because of miscalculation of weight
- Failure of planned demolition because unexpected construction details

3.7.4.1 Demolition of bridges

Extension or renewal of roads and railroads often require demolition of bridges, from small crossings over roads and railroads to major water crossings. In case of demolition of crossings over roads and railroads, the demolition of bridges must pay special attention to the traffic with respect to safety, as well as time of interruption of the traffic. Therefore, demolition of bridges must be planned in details according to the actual condition. Important railroad bridge and road bridge demolitions must be executed within a very short time limit during periods of minimal traffic intensity, typically between Saturday afternoon and Monday morning. The demolition work comprises the following activities:

- Preparatory work, removal of surfaces and earth filling, removal of non-critical items, such as lightning poles and installations, parapets and other work which do not influence the traffic below the bridge to be demolished. Also, cleansing of hazardous materials must be completed
- Closing the traffic and protection of the road/railroad under the bridge against falling debris
- Demolition of the bridge structure requires a total stoppage of traffic. The demolition of minor bridges, 2–4 spans, might be completed during one night
- Removal of debris and successive clearance and opening of traffic lanes
- Completion of the uncritical parts of the demolition work

The preferred methods for demolition of shorter crossings are breaking and crushing using heavy long-reach machines (see Figure 3.14). Considering time-critical demolitions, it is important to mobilize reserve capacity and prepare for the substitution of machines in case of machine fault or unforeseen problems. Sometimes, blasting might be applicable. The preparation of blasting with explosive charges in drill holes for blasting concrete, or placing charges on steel structures take much time, while the blasting itself and demolition of the structure takes seconds. Blasting is often rejected because of environmental impact considerations, as well as the need for security of the explosives with respect to risk, safety and criminal acts.

Figure 3.14 Demolition of a bridge crossing a highway starting on Saturday and opening all lanes Monday morning. (Photo: NIRAS.)

Demolition of long bridges, typically water crossings, are challenging jobs where many conditions need to be assessed, among others:

- Demolition methods including demolition of superstructure, road structure, pillars, foundations, towers, arches, etc.
- Environmental impact, including impact on fish and plants, pollution of water, noise, dust, vibration, special regard to rare species, etc.
- Navigation around the site
- Risk and security
- Health and safety
- Total economy

Spectacular blasting demolition of bridges have been seen in the United States, Europe and other places in the world. However, because of the risk of uncontrolled impact and damage to the marine environment, today many owners and authorities no longer consider the blasting method.

3.7.4.2 Demolition of the San Francisco–Oakland Bay Bridge

The San Francisco – Oakland Bay Bridge is one of the greatest American bridges. The San Francisco – Oakland Bay Bridge opened to traffic in 1936. The bridge is actually several structures with distinctly different systems, strung together to form a 13.7-km cross-bay roadway, with nearly 7.1 km over the water. The east crossing, 3,417 m, consisting of several different steel truss systems with two decks has been replaced by a new bridge. The east crossing was locally damaged during the Loma Prieta earthquake of 1989. A 15-m section of the top deck slipped off its support at an expansion joint; that end of the section then collapsed onto the lower deck and one motorist was killed (Middlebrook and Mladjov 2014).

According to information from the *California Department of Transportation* (Caltrans 2015), demolition of the nearly 2-mile long original east span has been taking place in three phases, with crews carefully

Figure 3.15 Demolition of the eastern crossing of the San Francisco–Oakland Bay Bridge, 2013–2017. Left: Shows the remaining part of the east crossing with the two piers in the foreground prepared for blasting. Right: Shows the work of dismantling the steel structure. (November 2015.)

dismantling the 77-year-old east span, section-by-section, in roughly the reverse order of how it was built in the mid-1930s.

- Phase 1 – Cantilever: Crews initially worked in the westward direction, toward Yerba Buena Island, taking apart the cantilever section and demolishing the S-Curve. Removed 11 June 2015.
- Phase 2 – Trusses: Included moving east to dismantle the truss spans that stretched east of the cantilever to the Oakland shore. Last truss span removed 28 March 2017 (see Figure 3.15).
- Phase 3 – Support piers: Entailed the demolition team heading to the waterline to remove the pier supports and pilings and then extracting the marine foundations that supported the original span down to the mud line. Project completed by the end of 2017.

When the entire demolition of the Old Bay Bridge is complete, over 58,000 tons of steel and 245,000 tons of concrete will have been removed.

The overall demolition method used was dismantling the superstructure cantilever and trusses by cutting the steel members and removing them by crane. The supporting piers were demolished by controlled blasting with a high degree of safety and environmental protection. To protect the marine environment, the piers were surrounded by an air carpet to absorb the emission of underwater shock wave energy.

3.7.4.3 Planned demolition of the Storstroems Bridge, Denmark

The bridge between two Danish islands, Sjaelland and Falster, which is 3.2 km long, was constructed in 1934–1937 with 50 spans, including three main spans (arches) made of 265,000 t of concrete and 20,000 t of steel (see Figure 3.16). The scale of this bridge is on par with the above-mentioned East Bay Bridge.

Figure 3.16 Storstoems Bridge, planned to be demolished in 2023.

Because of the degradation of the concrete deck and the rising traffic intensity, the Danish road directorate has decided construct a new bridge and demolish the old one. Construction of the new bridge started in 2018 and the demolition of the old bridge is expected to start in 2023.

Demolition strategy:

- Disassembling the superstructure by lifting each separate span using jacks and elevating work towers; transport on barges to down sizing site
- Demolition of substructure by blasting of piers and foundations. The concrete rubble is collected from the seabed and transported to the crushing plant, e.g., the nearby harbour. Alternatively, the rubble is left on the seabed.
- Foundation plates are left on the seabed.
- The south abutment is removed to ground level.
- The north abutment is preserved as a public vantage point.
- Clearance of the asbestos membrane in the concrete plates of the superstructure is discussed. It is proposed to mill the upper layer and separate the amount of clean concrete from the asbestos-contaminated concrete, maximising the amount of recyclable concrete.

The concrete of the superstructure, approximately 30,000 t will be disposed after demolition as contaminated waste. The concrete of the substructure, approximately 135,000 t, is expected to be clean and recycled locally for road constructions, or as aggregates in new concrete for low grade structures.

The blasting of the piers will be performed as controlled blasting with special care for the environment. Threshold values of peak pressure from underwater shock waves will be based on the risk assessment of impact to mammals and fish.

3.7.5 Developing methodologies

Research and development of methodologies based on known demolition technologies takes place currently. The Kajima Cut and Take Down Method of high-rise buildings deserves a presentation. Instead of demolition from the top, the Cut and Take Down Method operates with successive lowering of the building on jacks, and demolition of the lowest floor by machines on ground level. The demolition starts with supporting the building at the ground floor. After demolishing the ground floor structure, the building is lowered by one height of floor, and the next floor structure is then demolished. This process continues stepwise until the demolition of the building is complete.

The method requires a controlled system of jacks and a specially designed ground frame to support the building. For further information, please refer to Kajima Corporation's website: www. kajima.com.

3.7.6 Summary of demolition methodologies

The demolition methodologies presented above are summarized in Table 3.13.

Table 3.13 Brief characteristics of the most common demolition methodologies

Demolition task	Demolition technology	Comments
Buildings		
Height below 10 m (3 storages)	Excavator with bucket	
Height between 10 m and 30 m	Crane and ball, long-reach excavator	Crane and ball needs space
Height between 30 m and 50 m		Crane and ball need space. Blasting needs space in falling direction
Height above 50 m	Blasting, machines from the top, disassembling	Blasting needs space in falling direction, machines require load-bearing decks
Towers, chimneys, silos, etc.		
Tall structures, dense surroundings	Small machines from the top, dismantling	
Structures below 50 m	Crane and ball, long-reach excavator, blasting, machines from the top	Good space needed
Structures above 50 m	Blasting	Good space needed
Bridges		
1–4 span motor road bridges	Excavators with breakers or crushers	Blasting might be applicable. In case of time-critical demolition, contingency measures are needed
Multispan bridges	Superstructure: blasting, dismantling Substructure and foundations: Blasting	Combination of various demolition technologies

Chapter 4

Recycling

The building is not complete but it is always under construction – it is always both old and new.

Wang Shu, Chinese Architect

4.1 INTRODUCTION

Recycling of CDW has taken place all through time in human history. Stones, brick, metal and other imperishable materials for shelters against the climate have always been recycled without any specific need for research and development. Breaking rocks and producing stones in suitable sizes and shapes for construction of buildings, churches, fortresses, harbours and other structures has required huge manpower through the ages. The long distances of transport, including international maritime transport, has added value to the stones. Therefore, recycling of stones and bricks from old abandoned and destroyed building, particulary war-damaged structures, has always been economical feasible and aesthetically attractive. See examples in Figures 4.1–4.3. After World War II and during urban development in the industrial countries, up to the middle of the nineteenth century, recyclable materials have been used as they were found applicable on-site. From that time, higher demands and requirements for technical specifications undermined recycling of building materials, especially reuse and timber. Reuse of crushed concrete for other purposes than landfill was not relevant at that time (Figure 4.4).

Waste reduction, recycling and effective use of buildings and building materials are key issues in the circular economy and in CO_2 reduction. This chapter presents the output of selective demolition and state-of-the-art, opportunities and challenges for recycling of the most dominating building and construction materials, as presented in Figure 4.5. Mainstream recycling is the key issue, which means recycling of the inert materials such as concrete, masonry, stones, tiles and asphalt. From a global perspective, recycling of concrete is the most challenging development because of the

Figure 4.1 Reuse of building materials. Left: Building façade, functional and architectural expression. (Wang Chu, Chinese Architect, 2017) Right: Part of a wall, AD 100, Forum Romanum, Rome.

huge amounts of consumption of raw materials and various opportunities of recycling. Recycling of other materials such as scrap metal, gypsum and mineral wool is also presented in this chapter.

It should be noted that we are talking of recycling CDW into resources applicable for the building and construction industry as building materials. The following recycling issues are not presented:

- Recycled waste materials from other waste sectors to be reused in the building and construction, for instance recycling of plastic and paper for structural building materials
- Recycled CDW for other purposes, e.g., recycling of timber for furniture and shredding wood for gardening materials
- Recycled CDW for purely decorative purposes with no functional effect as building materials

4.2 RECYCLING OF CONCRETE

4.2.1 Global consumption of concrete

More than 30% of waste generated per year in the world is building waste; more than 50% of all CDW is concrete. According to Metha (Metha 2002), the global concrete industry uses approximately 10 billion tons of sand and rock each year, and more than 1 billion tons of CDW are generated every year. In principle, all concrete waste is inert and can be recycled. However, some concrete waste is polluted with harmful substances from paint and additives. At least 90% of concrete waste has a potential for recycling. Full implementation of recycling of concrete waste all over the world can save around 1 billion tons of natural resources. According to Vazquez (Vazquez 2012), some authors estimate that the replacement percentage of prime resources amounts to 15%. The recycling of this waste was heavily promoted in the 1970s and 1980s in several countries, on the premise of saving resources and land by slowly

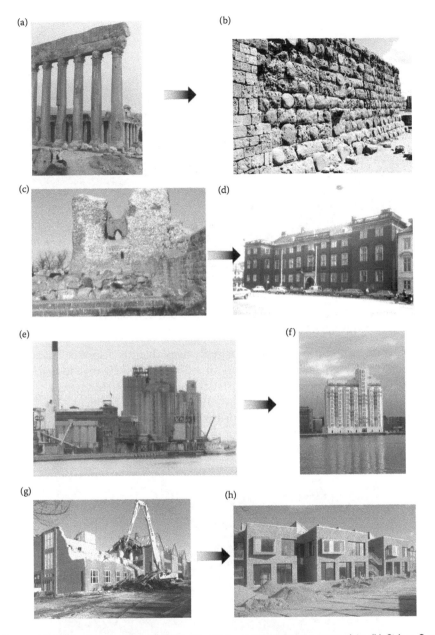

Figure 4.2 Historic recycling. (a) and (b): Old granite columns reused in (b) Sidon Sea Castle, Lebanon, built by the crusaders in the thirteenth century. (c) and (d): Bricks from old Kalø Castle, Jutland, Denmark, reused in a museum Charlottenborg, Copenhagen, seventeenth century. (e) and (f): Silos from an old factory reused for new dwellings. (Copenhagen, 2000.) (g) and (h): Bricks from an old building of Bispebjerg Hospital reused in façade of new school buildings, Katrinedals School. (Copenhagen 2016.)

Figure 4.3 Recycling of bricks and timber during urban development in Copenhagen in 1942. (Copenhagen City Museum.)

reducing the amount of landfills normally available for disposal. Since then, recycling of CDW has developed into an ambitious and creative industry that produces recycled aggregate for construction (Vazquez 2012).

Referring to the report prepared by Bio Intelligence Services, Acadis and the Institute for European Policy for the EU Commission in 2011, Bio et al. (2011) selected figures related to consumption and recycling of concrete, mainly in EU member states, which are listed in Box 4.1.

Figure 4.4 Recycling of crushed bricks in building foundations has been approved and has been commonly used through time.

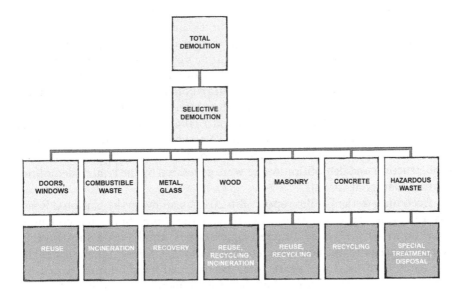

Figure 4.5 Typical output of selective demolition.

BOX 4.1 CONSUMPTION AND RECYCLING OF CONCRETE. SELECTED FIGURES

- Worldwide consumption of concrete, 2006: 21,000–31,000 Mt
- Total EU production of ready-mixed concrete, 2008: 900 Mt
- World production of concrete waste more than 1,000 Mt per year.
- More than 50% of all CDW is concrete. In the EU 50%–70%
- More than 90% of CDW concrete is recyclable
- EU average consumption of concrete per capita, 2006–2008: 0.80 m³
- The ready-mixed concrete industry alone consumed 216.2 Mt cement in 2008, about 75% of the EU-27 production
- Production of concrete in the EU-27, 2008: 1,350 Mt, including 900 Mt ready-mixed concrete, and 200–250 Mt precast concrete
- Waste concrete generation in EU-27, 60%–70% of the total C and D waste (320–380 Mt in 2008)
- Potential recycling rates: 20% of coarse aggregates used for production of structural concrete can be made of recycled aggregates.

Source: Bio Intelligence Service, Acadis, Institute for European Environmental Policy, Service contract on management of construction and demolition waste – SRI, Final report Task 2, February 2011.

Jianzhuang Xiao et al. report (2012) that China is a large resource consumer with a shortage of resources and extensive management; its economic growth is increasingly constrained by resources and the environment. According to statistics, China's consumption of cement is 820 million tons and accounts for 55% of the world demand. It is estimated that approximately 200 million tons of waste of concrete are currently produced annually in the mainland of China. Moreover, some natural disasters such as the *Wenchuan earthquake* (2008) and *Yushu earthquake* (2010) in China have resulted in a great quantity of waste concrete. Additionally, in the construction process of the *Shanghai Expo 2010*, nearly 300 million tons of construction and demolition waste were generated by the related demolition work (Xiao et al. 2012).

According to the EU Bio report (Bio 2011), a study by the National Ready Mixed Concrete Association (NRMCA) in the United States has concluded that up to 10% recycled concrete aggregate is suitable as a substitute for virgin aggregates for most concrete applications. UK research indicates that up to 20% of total aggregates may be replaced by good quality crushed concrete. Under these conditions, recycled concrete can be used for most common concrete applications. Actual practices vary greatly; for example, countries like the United Kingdom and The Netherlands already achieve a recycled concrete content of 20%, whereas this application is almost nonexistent in other countries such as Spain.

4.2.2 Research and development

Since Agenda 21, the Rio Declaration on Environment and Development was launched in 1992, and sustainable development has been one of the key issues of modern society. Recycling of concrete has been an important issue in the United States, Europe and other industrial parts of the world. Many years ago – in the 1970s – the US Federal Highway Administration (FHWA) accepted recycling of concrete pavement with recycled crushed materials as aggregates in new concrete (Recycled Aggregate Concrete [RAC]).[1] In the 1990s, the American Concrete Institute (ACI) realized that, even if concrete is an environmentally friendly material, Portland cement is the critical component of modern-day concrete. To address this issue and the relationship between sustainable development and concrete technology, the ACI Board of Directors, in 2000, formed a Task Group on Sustainable Development and Concrete Technology. Its mission was to encourage the development and application of environmentally friendly sustainable concrete materials, designs and constructions. One of the most important issues of sustainability was the use of recycled aggregates.

---

[1] In this book, Recycled Aggregate Concrete (RAC) means bound recycling of concrete, crushed concrete used as aggregates in new concrete. Recycled Concrete Aggregate (RCA) means unbound crushed concrete and recycling of concrete as unbound materials.

Fortunately, some ACI members had been far-sighted enough in 1985 to organize Committee 555 – Concrete with Recycled Materials. In 2001, the committee submitted a report *Removal and Reuse of Hardened Concrete* (ACI 2001) followed by the report *Recycling Concrete and Other Materials for Sustainable Development* in 2004 (ACI 2004). The two reports have established a very good basis for the future work of the ACI on the sustainability and recycling of concrete.

Parallel to the work of FHWA and ACI in the United States, the recycling of CDW has emerged as a socio-economic priority within the European Union. A considerable amount of research and development has taken place in the frame of RILEM.[2] From the late 1970s, recycling of building materials were introduced in international research and development. In 1981, Prof. Torben C. Hansen at the Danish Technical University reorganized the RILEM Technical committee 37-DRC on Demolition and Reuse of Concrete with members from Belgium, Denmark, France, Germany, Great Britain, Japan, The Netherlands, Sweden and the United States. The task of the committee was: (1) To study the demolition techniques used for plain, reinforced concrete, and pre-stressed concrete and to consider developments in techniques, and (2) to study technical aspects associated with reuse of concrete and to consider economic, social and environmental aspects of demolition techniques and reuse of concrete. The work of TC37-DRC was concluded with an international conference in Tokyo, Japan 1988 (Kasai 1988). This committee was followed in 1988 by the Technical Committee TC-121-DRG on *Guidance for Demolition and Reuse of Concrete and Masonry* with the main objective of preparing draft guidelines for demolition and reuse of concrete and masonry with special regard to urban development and clearance of urban areas after major natural disasters and wars. The results of TC-121-DRG were presented at the third international RILEM symposium on *Demolition and Reuse of Concrete and Masonry* held in Odense, Denmark 1993 (Lauritzen 1993). In 2000, the state-of-the-art-report of RILEM Technical Committee 165-SRM on *Sustainable Raw Materials* was edited by C.F. Hendriks and H.S. Pietersen (Hendriks et al. 2000) followed by TC 198-URM on the *Use of Recycled Materials in Buildings and Structures,* and a worldwide conference on recycled materials in 2004 in Barcelona (Vazquez et al. 2004).

As mentioned in the preface of the proceedings (Vazquez et al. 2004), the knowledge in this field has strongly increased: technical performance of recycled materials is used for many purposes such as environment and health, economics, processes for cleansing, upgrading and different type of production methods. In addition, legislation and specification in many countries have made it possible for the quantities of recycled materials to

---

[2] RILEM: Réunion Internationale des Laboratoires d'Essais et de Recherches sur les Matériaux et les Constructions/International Union on Testing and Research Laboratories for Materials and Structures

increase strongly. In 2012, the RILEM TC-217 followed up the technical development on demolition and reuse of building materials, publishing a report *Progress in Recycling in the Built Environment,* edited by Prof. Dr. Enric Vazquez (Vazquez 2012). The objective of the publication is to collect the data on advancements, achievements, problems and solutions most representative of the various situations. Thereby, each country can learn from examples and underpin a pedagogy for the effective implementation of recycling. The report is the result of the work of the RILEM technical committees and contributions of the Symposium of Sao Paulo 2008, the International RILEM conference on *Progress of Recycling in the Built Environment* in Sao Paulo, 2009, and the International RILEM Conference on *Waste Engineering and Management* in Shanghai, 2010. Following the RILEM tradition, the second objective of this publication is to collect and report on the advances in concrete recycling produced since 2005.

The research and development work on recycling concrete is followed up in RILEM by the TC-RAC on *Structural Behaviour and Innovation of Recycled Aggregate Concrete (RAC).* It is expected that the TC-RAC will publish a report on recycled aggregate concrete in 2017, and a report on RAC structural performance in 2018.

The *Handbook of Recycled Concrete and Demolition Waste* by F. Pachero-Torgal et al. (2013) presents a very comprehensive and detailed guidance on recycling concrete. The handbook comprises updated state-of-the-art including the international research and development mentioned above and is very useful in practice.

In spite of a lot of fundamental research and development on recycling concrete taking place throughout the past 40 years, recycling concrete and substituting natural aggregates with crushed aggregates in new concrete has not been a success in the building and construction industry. In North America, a research has been conducted on various aspects of recycling concrete. However, despite the many favourable research findings regarding the good short- and long-term performance, crushed concrete is still predominantly used as backfill, base and sub-base material rather than substitute for virgin aggregate in structural quality concrete (Vanderlev 2012). In Europe, the EU and some member states (Germany, the United Kingdom, The Netherlands, Denmark and others) encourage the building and construction industry to promote the recycling of concrete.

4.2.3 Application of recycled concrete

Concrete is made from coarse aggregates (gravel or crushed stone), fine aggregates (sand), water, cement and admixtures. Cement is a hydraulic binder that hardens when water is added and represents between 6% and 15% of the concrete mix depending on the application, while aggregates typically represent some 80% in mass. Concrete waste is the major fraction of CDW, and recycled concrete waste has a high variety of applications.

Table 4.1 Hierarchy of recycling concrete

Ranking	Reuse, recycling, recovering
1	Reuse of concrete structures
2	Reuse of concrete construction elements
3	Recycled Aggregate Concrete (RAC) Recycling of concrete as aggregate in new concrete)
4	Recycled Concrete Aggregate (RCA) Recycling of concrete as unbounded road materials
5	Recovered crushed concrete as backfill

Note: Ranking the value of applications for recycling concrete.

From a hierarchy perspective of the value of the recycled material, the general applications are ranked as presented in Table 4.1.

Concrete produced using recycled aggregate, or combinations of recycled aggregates and other aggregates, is called *Recycled Aggregate Concrete (RAC)* – aggregates in new concrete. Aggregates produced by crushing of concrete is called *Recycled Concrete Aggregate (RCA)* – unbounded aggregate for base materials and bounded aggregates for new concrete (Hansen 1992). The benefit of RAC is saving the aggregate for the new concrete. Cement, sand and water must be provided. The benefit of RCA is saving aggregates for unbound sub-base material and to provide up to 100% substitution of natural (primary) materials.

4.2.4 Reuse of buildings and concrete structures

Old industrial concrete buildings, for instance silos and factory buildings are often retained and converted to other purposes, such as dwellings or offices. The economic advantage is keeping the concrete structure and the height of the building. Usually, much work with transformation of the building, establishing openings, and so on, must be expected.

Reuse of buildings for other purposes is usually the most advantageous recycling mode. In many industrialized cities, concrete structures such as towers, silos, and industrial plants are seen reused as offices or dwellings. The opportunity for reuse of buildings depends on various factors and decisions with respect to function, architecture, economy, local policies, and so on (see Chapter 5).

4.2.5 Reuse of concrete construction elements

Since the middle of the nineteenth century, concrete has been commonly used to construct buildings with pre-fabricated concrete building elements. When a precast concrete building is demolished, contractors would prefer to sell the entire structure for reconstruction as an entity, or to utilize the whole precast units for other construction purposes. Presupposing the elements could be separated during demolition of the building, these elements could be reused. However, practical experiences in Denmark have shown that many technical and legal implications prevent the reuse of construction elements. First, concrete elements in building structures are normally connected with locked joints of reinforcement bars and sealed with concrete mortar. This makes it very difficult to separate the elements without damaging them. Second, there are problems associated with the reuse of precast concrete units in new buildings. The elements must be handled and reused in an applicable way that makes sense. The reused elements must fit into the new building structure with respect to the strength, geometry and dimension of the elements. Third, it is problematic to get the proper documentation about the quality of the used elements. Fourth, responsibility for the structural integrity of the new building has legal implications, as there is a question of product liability, including the aspects of legal responsibility for the reuse processes. It might be very difficult to get the structure insured, because the experiences in the field of use of recycled materials in the construction industry are very poor.

4.2.6 Design for disassembly

Looking into the future, we might break the ground for reuse of construction elements by *designing the structures for disassembly*. With respect to the requirements of sustainable construction, green change and the circular economy, the design of concrete structures for disassembly is an important concept. A Danish team of specialists headed by the architect firm 3XN, and its research and development subsidiary GXN, and the Danish contractor MT Højgaard have developed solutions for the design of concrete structures for disassembly (Guldager Jensen and Sommer 2016).

According to the report of GXN and MT Højgaard, five principles should be considered when designing for disassembly:

- *Materials*: Choose materials with properties that ensure they can be reused
- *Service life*: Design the building with the whole lifetime of the building in mind
- *Standards*: Design a simple building that fits into a larger context system
- *Connections*: Choose reversible connections that can tolerate repeated assembly and disassembly
- *Deconstruction*: As well as creating a plan for construction. Design the building for deconstruction

The implementation of design for disassembly follows the sustainable building certification according to the official certification schemes: DGNB, BREAM, LEED and others. Today, the sustainable certification and the design for disassembly is voluntary. The public requirement of design of buildings for disassembly needs legal authority. This might take many years to accomplish. The mentioned principles of design for disassembly deal with construction today and in the future, aiming at the end of the buildings' life cycles. The principle of design for disassembly is applicable for design of interim structures, for instance buildings and structures in major constructions projects. The design for disassembly will be presented and discussed in Chapter 7.

4.2.7 Recycling of crushed concrete as aggregates in new concrete (RAC)

4.2.7.1 State-of-the-art

Crushing of hardened concrete into secondary raw materials into aggregate for production of new concrete is considered as a higher quality of recycling than the recycling of unbound crushed concrete. Prompted by earlier successful applications in the United States, a number of experimental projects were carried out in Denmark in 1986–1995, which showed that it was possible to use size-fractions of demolished and crushed concrete larger than 4 mm as aggregate in the production of new concrete for passive and moderate environments. Based on the results of these projects, specifications for recycled concrete in passive and moderate environments have been prepared both nationally in Denmark (Danish Concrete Association 1989), and internationally by RILEM TC-DRG-121 (RILEM 1994). The specifications concern concrete produced with the 4/32 mm recycled aggregate fraction only. If it is desired to use the 0/4 (up to 4 mm) fraction also, the propriety of doing so must be demonstrated in each individual case, which is difficult and expensive. The compressive strength of concrete produced with 4/32 mm recycled aggregate is somewhat less than the strength of regular concrete of the same consistency produced with the same cement content. The modulus of elasticity is also somewhat smaller, while creep and shrinkage are higher than for regular concrete. However, this can be taken into account by application of the above-mentioned specifications. It has been shown in the United States that high-quality concrete exposed to severe environments, for example in pavements, can be produced from demolished high-quality concrete. However, it is not possible to produce concrete for severe environments from crushed old concrete of low quality. Crushed old concrete that is of low strength and not frost resistant, or which is produced from aggregates prone to alkali-silica reactions, should not be used for the production of new concrete exposed to severe environments.

In Europe, the use of RAC is less commonly accepted, but almost every application found is practiced in more than one country. The major differences among the countries appear in the technical and environmental requirements for the use of recycled aggregates. Limit values are often

Table 4.2 Maximum percentage of replacement of coarse aggregates with recycled aggregate (% by mass), according to EN 206:2013

Recycled aggregate (RAC) type	Exposure classes			
	XO	XC1, XC2	XC3, XC4, XF1, XA1, XD1	All other exposure classes
Type A, Known source	50%	30%	30%	0%
Type B, Other sources	50%	20%	0%	0%

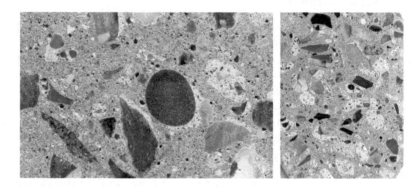

Figure 4.6 Cut sections of two types of RAC. The figures show crushed aggregates comprising old stones with layers of old cement pasta and particles of crushed cement pasta. (PELCON.)

based on national legislation. Though national legislation and standards are replaced more and more by European standards, the interpretation of these standards is different in each country.

Referring to the European standards *EN 12620:2013 Aggregates for Concrete* and *EN 206:2013 Concrete-Specification, Performance, Production and Conformity,* RAC can be used in new concrete with a maximum percentage of replacement of coarse aggregates as shown in Table 4.2. However, higher percentages up to 100% might be applicable.

It is important to know that RAC has other properties, which must be taken into account when using RAC in the construction industry. Compared to concrete made by primary aggregates, the most important characteristics of RAC are the following (Hansen 1992; IRMA 2006):

- Recycled aggregates *contain old mortar* from the original concrete. The volume percentage depending on the fraction range from 25% for 16–32 mm, to 60% for 4–8 mm (see Figure 4.6)

- The *density* of the recycled aggregates depends on the quality of the original concrete. It is expected that the density is approximately 5%–6% lower than the density of the original concrete due to the relatively low density of the old mortar
- Generally, the *compressive strength* of concrete with recycled aggregates is lower than similar corresponding concrete with primary aggregates (see Table 4.3)
- A mandatory precondition for recycling aggregates is that the materials are clean without content of *hazardous materials* above the approved threshold levels
- Due to the large amount of old mortar with a comparative low *modulus of elasticity*, which is attached to the original particles in recycled aggregates, the modulus of elasticity of recycled aggregates is always lower than that of corresponding control concrete with primary aggregates
- Because of the higher content of mortar, recycled aggregate concrete has *higher creep* compared to the corresponding concrete with natural aggregates. The results of research have shown 20%–60%, depending on the strength of the new concrete
- Drying *shrinkage* of RAC made by coarse recycled aggregate and primary sand is approximately 50% higher than shrinkage of corresponding concrete made with primary aggregates
- There is evidence that RAC might be problematic with respect to *frost resistance*. However, studies on frost resistance are still lacking, and it is thus recommended to take special care of the frost resistance problem in relevant construction scenarios

Xiao et al. (2012) have presented a state-of-the-art report on the relevant research and findings on the mechanical properties and structural

Table 4.3 Compressive strength of original concrete and recycled aggregate concrete made from the same original concrete using recycled coarse aggregate and various proportions of recycled fine aggregate and natural sand

	Compressive strength of concrete (MPa)			
w/c	Natural coarse and fine aggregate (original concrete)	Recycled coarse aggregate and 100% natural sand	Recycled coarse aggregate, 50% recycled fine aggregate, and 50% natural sand	Recycled coarse and 100% recycled fine aggregate
0.45	37.5	37.0	34.0	30.0
0.55	28.9	28.5	25.0	21.5
0.60	22.0	21.0	17.5	13.0

Source: Hansen, T. C., Recycling of demolished concrete and masonry, RILEM report 6, E & FRN SPON 1992.

Note: The concrete strength is related to the water/cement ratio (w/c).

performance of RAC in mainland China. The main conclusions, which are in line with above-mentioned characteristics, are the following:

- The compressive strength of RAC relies heavily on the water/cement ratio
- The quality of RCA has a considerable effect on the probability distribution for the strength of RAC
- The compressive, tensile and shear strength of RAC are generally lower than that of conventional concrete
- The modulus of elasticity for RAC generally reduces as RCA increases; however, the strain at peak stress is larger than that of conventional concrete
- The RCA replacement percentage has nearly no influence on the bond strength between RAC and deformed rebar
- The RCA has a remarkable influence on the residual strength of RAC after high temperatures
- The carbonation resistance of recycled concrete decreases with the increase in the replacement rate of recycled aggregate, and the anti-carbonizing capability of recycled aggregate is worse than that of normal concrete
- The chloride resistance of recycled concrete reduces with the increase in replacement rate of recycled aggregate while fly ash can improve the chloride resistance of recycled concrete; the best quantity of fly ash is from 10% to 15%. Recycled concrete has good resistance to chloride penetration, but its resistance to chloride penetration is somewhat less than for normal concrete
- Recycled concrete absorbs more water than normal concrete and harmfully influence frost resistance
- The replacement rate of recycled aggregate has the greatest impact on the abrasion resistance of recycled concrete and its abrasion resistance is lower than for normal concrete
- The structural behaviour of RAC elements/members is generally weaker in comparison to that of structures made of natural aggregate concrete (NAC)
- Through proper design, RAC can be used as a structural material from the view point of the loading capacity behaviour
- More tests on the serviceability properties of RAC elements and structures need to be carried out in the future
- The reliability index of recycled concrete beams is proved to meet the requirements of the *Unified Standard for Reliability Design of Building Structures* (GB 50068-2001)

There are major similarities as well as major differences among the European countries in the way they deal with concrete products from CDW. The similarities are found in the applications. Almost all European

Table 4.4 Overview of relevant European standards regarding the use of recycled
aggregate concrete (RAC)

EN 206:2013 Concrete-Specification, performance, production and conformity
EN 12620, 4. edition, 2013–05 Aggregates for concrete
EN 932, part 3 Petrographic description of the aggregate
EN 933, part 11 Classification test for the constituents of coarse recycled aggregate
EN 1097, part 2 Determination of the Los Angeles coefficient
EN 1097, part 6 Determination of particle density and water absorption
EN 1367, part 1 Determination of resistance to freezing and thawing
EN 1744, part 5 Determination of acid soluble chloride salt

countries use recycled aggregates for road-base and foundation layers. Other applications, such as the use as gravel substitute in concrete are less commonly accepted but almost every application found is practiced in more than one country. The major differences among the countries appear in the technical and environmental requirements for the use of recycled aggregates. Limit values are often based on national legislation. Though national legislation and standards are increasingly replaced by European standards, CEN,[3] the interpretation of these standards is different in each country. An overview of relevant European standards regarding the use of recycled aggregates is presented in Table 4.4. These standards all apply to both primary aggregates and to recycled aggregates. Recycled aggregates should comply with the same demands for the same application as primary aggregates.

All aggregates that are brought to the economic market should be provided with a CE marking based on the harmonized European standards. A CE-label indicates that products are tested according to European standards and that the performance/features of these products are declared. CE-marking does not require that the demands be identical in all European countries, but it requires that the declaration of the product properties is harmonized: the same tests are applicable everywhere. In practice it appears that the CE-marking is not yet fully operational.

4.2.7.2 Practical experiences

Hansen (Hansen 1992) reports practical case stories including many successful projects in United States on recycling road concrete pavements in several states conducted by the Federal Highway Administration (FHWA) in the 1970s and 1980s. In 2002, the FHWA issued the *Formal Policy on the Use of Recycled Materials* (FHWA 2002). According to the policy, the FHWA has established agency goals for enhancing the human and natural environment, increasing mobility, raising productivity, improving safety throughout the highway industry and preserving national security. All of these goals are stated in the FHWA strategic plan ensuring that the FHWA

---

[3] European Committee for Standardization; Comité Européen de normalisation, CEN.

recycling policy and recycling programs can offer engineering, economic and environmental benefits:

1. Recycled materials should get first consideration in materials selection
2. Determination of the use of recycled materials should include an initial review of engineering and environmental suitability
3. An assessment of economic benefits should follow in the selection process
4. Restrictions that prohibit the use of recycled materials without technical basis should be removed from specifications

The technical requirements for the recycled concrete materials are given in Technical Advisory T 5040.37 (FHWA 2007) comprising the description of processes of recycling and requirements for ensuring the quality of the RCA aggregates.

Apart from FHWA recycling concrete pavement, recycling of concrete as aggregate in new concrete is not commonly used in United States.

In Europe, most road and highway pavements today are made of asphalt. Therefore, recycling of concrete pavements is not commonly used. Recycling of unbound recycled concrete in road base is the priority of recycling concrete debris in most EU states. As mentioned before, even though technical documentations and standards open opportunities for using RAC, there is very little interest in RAC technologies. However, referring to Vazques (Vazquez 2012), some EU member states have reported their experiences.

Anette Mueller, Germany (Mueller 2012), reports that increased efforts from time to time can be observed to establish closed circuits in the building sector focusing on reuse of recycled aggregates for concrete production. The first demonstration activity was the use of recycled aggregates for the construction of an office building of the Deutsche Bundesstiftung Umwelt (DBU) in Osnabrueck in 1994. Then a number of buildings followed in connection with the research project *Baustoffkreislauf im Massivbau BiM*. Recently, concrete was made of RC aggregate according to the current standards. It was used in the context of several flagship projects in residential and administrative buildings. In 2017, CEMEX Germany informed (CEMEX 2017) on the development of RAC structural concrete for a new building at the Humboldt University in Berlin. Approximately 3,800 cubic metres of RAC, Class C30/37, with recycled aggregates fraction 8/16 mm were used.

According to Isabel Martins (Martins 2012), Portugal has implemented the following guides:

- Guide for the use of coarse recycled aggregates in concrete
- Guide for the production of recycled hot mix asphalt
- Guide for the use of recycled aggregates in unbound pavement layers
- Guide for the use of recycled materials coming from construction and demolition waste in embankment and capping layer of transport infrastructures

Figure 4.7 Production of recycled aggregates, Theo Pouw, Utrecht. (From Theo Pouw Group, www.theopouw.nl. Company information, last visited: 2017.)

Case stories on construction of a new pier of Lisbon Airport (2008–2011) and other recycling projects are reported (Martins 2012).

The Netherlands is the spearhead of recycling. Referring to several publications and proceedings, Prof. Charles F. Hendriks, Technical University Delft, has contributed to research and development of recycling concrete, not only in The Netherlands but also on a European and international level. The publication *The Building Cycle* (Hendriks et al. 2000) presents a comprehensive guide on recycling comprising processes from demolition to the use of recycled materials. The guide includes the term *Chain Management* that is the key concept of *Circular Economy* today.

From January 1985 it has been allowed to use recycled aggregates for production of concrete in The Netherlands (Hansen 1992). Today, it is the policy of the Dutch government to encourage the construction industry to substitute up to 20% natural aggregates with RCA. The MIA/Vamil[4] 2017 programme supports production of concrete substituting natural aggregates with at minimum 30% RCA with 50 EURO per cubic meter concrete (Theo Pouw 2017) (see Figure 4.7). The policies of the Dutch government have been a major driver for a specialized industry producing high-quality recycled aggregates for concrete production in The Netherlands.

Referring to the work of RILEM TC121-DRG and the development of RILEM recommendation *Specification for concrete with recycled aggregates* (RILEM 1994), several demonstration projects were conducted in Denmark in the 1980s–1990s on recycling crushed concrete as aggregate in concrete foundations. In 1989, two old concrete railway bridges were demolished and the concrete was crushed and screened into a 4/16 mm and a 16/32 mm fraction. 400 t of recycled aggregate was produced and used in the construction of a residential building and a parking garage. The fresh concrete was produced and delivered as planned, and the properties of the hardened concrete were satisfactory. However, high-water absorption of the recycled aggregate required careful control of the batching and mixing

---

[4] The Dutch Milieu-investeringsaftrek (MIA) en de Willekeurige afschrijving milieu-investeringen (Vamil), Netherlands Enterprise Agency.

process. There were also practical problems of an operational and logistics nature (Lauritzen and Hansen 1997).

It was concluded that a viable production of recycled concrete aggregate of high quality is possible only at a plant ensured a yearly production of at least 100,000 t. This is due to the fact that the plant must comprise special crushing and transport facilities. The parties involved must also be familiar with the special routines and working procedures, which are associated with selective demolition and handling of recycled concrete. The project was followed by three apartment houses built in 1991–1994, predominantly with recycled materials, including concrete with aggregates of crushed concrete. The three projects are described in Section 4.9.

After 20 years of stagnation in the RAC area because of political, technical and pollution barriers, new initiatives on promoting RAC in buildings and structures are being seen in Denmark. A new Pelican Self-storage warehouse has been constructed in Copenhagen with foundations of concrete with crushed materials from older buildings on the site (see Figure 4.8).

The economic and environmental benefit in recycling concrete depends on the opportunities of recycling on-site. That means, the demolition, crushing of concrete, and production of new concrete should take place on-site in order to save transport, energy and CO_2. Today, the building and construction industry in European urban areas prefers ready-mix concrete from stationary concrete plants delivered by concrete trucks to the construction site. On-site concrete mixing is normally only preferred in case of major infrastructure projects. Therefore, the major challenges of recycling concrete are of logistic nature. The optimal conditions are on-site crushing, mixing and curing, which need careful planning and space. The benefits of using RAC concrete, compared with traditional ready-mixed concrete, can be calculated by LCA analysis. We will discuss this issue in Chapter 5.

Research and development activities have focused on structural high-quality RAC. It is important to notice that more than half of the production of concrete is non-classified and low exposure class with a high potential of substituting with RAC (see example in Figure 4.9).

4.2.8 Recycling of concrete as unbound road-base materials (RCA)

Recycling of crushed hardened concrete into secondary raw materials for use as unbound road bases, sub-bases or fill is generally used all over the world. However, the quality of the materials varies from unsorted debris tipped off on the planned road to serious design and use of sorted materials in sub-base layers and top layers, as presented in Figure 4.10.

Depending on the particle size, recycled aggregates can function as a substitute for natural sand or stone under road constructions, pavements or buildings. Aggregates used for road base are mostly applied in layers. The total thickness is often limited to several decimetres. Aggregates used as

Figure 4.8 Pelican self-storage house in Copenhagen, 2016–2017. Top: Shows the mobile mixing plant. Bottom left: Shows the concreting. Bottom right: Shows the final results of a concrete wall.

Figure 4.9 Module concrete blocks made of RAC for dividing walls at Theo Pouw's plant for recycling aggregates.

foundation layer under building are often used in thicker layers. By application in thick layers, the aggregates might be used for multiple purposes: levelling and elevation as well as foundation layer. Aggregates used for road base or foundation must be resistant to heavy loads. Therefore, requirements regarding grading and compression should be met. Like aggregates used for filling, the aggregates used for road base and foundation layers also become part of the subsoil, so environmental aspects should be paid attention to as well.

Research and development work associated with the crushing of demolished concrete into secondary raw materials has a long tradition in Denmark. It began in 1983 with the crushing and recycling of 150,000 tons of demolished concrete from the old runway at the Copenhagen International Airport, as unbound base course material in connection with reconstruction of the main runway. The results were good. The base course

Figure 4.10 Temporary road made of crushed concrete, UN City in Copenhagen.

material, which was produced from the old concrete, was of better quality than the corresponding material produced from natural sand and gravel. The modulus of elasticity of the recycled materials was 38% higher than that of natural materials. This allowed a reduction of the thickness of the asphalt surface layer, thus improving the pavement economy by 20%–40% compared to traditional construction. At the same time, the environment was spared thousands of truckloads to and from the airport and the disposal of 150,000 tons of old concrete as well as a corresponding supply of new natural raw materials (Lauritzen 1997).

At a different location in Denmark, a 160 m long road in a residential district was built using crushed concrete in the fractions 0–32 mm and 32–55 mm as unbound base course material. The results of the project showed that crushed concrete can easily be placed, and that it has very good drainage properties compared to natural materials. On the other hand, crushed concrete serves poorly as capillary break. Therefore, it should not be used as subgrade material.

Danish research and development work has shown that by using a known crushing technology it is possible to produce a well-graded road base material from crushed concrete. However, due to practical circumstances it has become common practice in Denmark to use a mixture of 50% crushed asphalt and 50% crushed concrete rather than pure crushed concrete as unbound base course material in road construction (Lauritzen 1997). Practical developments of recycled road base materials, which have taken place in Denmark, are interesting and deserve further consideration. It was originally foreseen by the authorities that most recycled concrete should be used as aggregate for new concrete. This would require concrete waste received by recycling plants to be comparatively clean, and that crushed concrete could be screened into at least two size-fractions: a high-grade coarse material above 4 mm and a very much lower-grade fine material below 4 mm. According to current specifications, the propriety of using the fine fraction as concrete aggregate must be demonstrated for each individual project. However, screening of coarse concrete aggregate, and documentation of the quality of crushed concrete fines is expensive. These are some of the barriers to recycled aggregate for new concrete.

Instead, recyclers have concentrated their efforts on production of two different mixed materials for road base purposes. One is a blend of nominally 45% 0/32 mm crushed concrete and 55% 0/32 mm crushed bricks. The other is a blend of nominally 50% 0/32 mm crushed concrete and 50% 0/32 mm crushed asphalt. Both materials have the advantage of using typical demolition rubble, crushed as received by the plant to a maximum size of 32 mm, without further screening into two or more size-fractions, which inevitably would produce more material in the undesirable 0/4 mm fines fraction. From technical points of view it appears that mixed road construction materials are better suited for unbound base course purposes than clean fractions of concrete, bricks or asphalt. The mixture of crushed concrete and bricks has high strength, it can be used for unbound base course purposes even under very wet conditions. It offers good support for temporary access roads and its technical properties and use are covered by existing standard specifications.

The mixture of concrete and asphalt is in even higher demand by contractors for road base purposes. The lubricating properties of asphalt reduces the physical effort required to compact the material by rolling. The mixed granular material is stronger than compacted crushed concrete because of the binding properties of asphalt, and it supports static loads better than clean crushed asphalt because the binder material does not constitute a continuous matrix in the composite base-coarse material.

On the basis of practical experiences and further studies, it should be attempted to determine the limits to the proportions in which such mixtures of crushed concrete and crushed asphalt can be used as road base and sub-base materials. It is evident that the wider the spectrum of acceptable mix proportions, the less expensive will be the processing, and the lower will be the costs of the recycled products. However, it is only on the basis of comprehensive and systematic studies and practical experiences that norms and specifications for mixed recycled materials can be prepared. It should also be studied if blends of crushed concrete, bricks and crushed asphalt can be used as road base and sub-base materials, and how sensitive the properties of such blends are to changes in the mix proportions.

4.2.9 Crushed concrete as fill material

Crushed concrete and masonry products are used as an alternative for soil or natural stones in earthworks and elevations, the filling of holes, mines, canals, and so on, and at levelling terrains for building activities. Aggregates for filling can differ in size, source material and composition. The technical quality of the material is considered less critical as for use in other applications, since this application is mostly seen as a low standard application. Environmental aspects, however, can be very relevant since the aggregates become part of the subsoil and might influence the quality of the surrounding soil and groundwater as a result of leaching.

4.2.10 Production of recycled aggregates

In principle, production of recycled aggregates is similar to the production of primary aggregates. The typical production plant consists of a crusher and a system of bands and sieves, depending on the required quality.

As distinct from production plants for primary aggregates, secondary aggregates are normally produced on mobile crushers. In major demolition jobs with good space and opportunities to use recycled materials on-site, the crushing plants should be established on the demolition site, or on a site close to the demolition site and the site for the use of the aggregates, depending on the logistics and economic preferences (see Chapter 5).

The most simple production plant is a mobile crusher for crushing aggregates with no specific requirement other than the size, for instance 0–32 mm material applicable for fill materials. For the production of aggregates qualified for RAC, the fine fractions 0–4 mm must be sorted out, and the production materials must be screened in two or three fractions in order to fulfil the requirements of particle distribution in accordance to the given screen curve (Figures 4.11 and 4.12).

Three types of crushers are used in the raw materials business (see Figure 4.13): jaw crusher, compact crusher and cone crusher. The jaw crushers is the preferred crusher type in the demolition and recycling industry because of its all-round capacity with respect to sizes of debris, types of debris stones and reinforced concrete and maintenance. A medium type crusher has a capacity of 200–300 t of concrete debris with a maximum size of 600–1,000 mm. The design criteria of a recycling plant depends on the logistic setup and demand of the product.

The main characteristics of the three types of crushers are as follows (Alaejos et al. 2012):

- Impact crushers give RA a more rounded shape, which is beneficial for engineering performance
- Impact crushers provide aggregates with very good grading
- Impact crushers require more maintenance as they suffer more wearing than the jaw crushers
- Jaw crushers have longer service life, as they suffer less wearing
- Jaw crushers admit a size block of up to 1,500 mm but do not reduce the maximum size under 60 mm

- The efficiency of jaw crushers is better, as they just reduce the particle size
- Jaw crushers give aggregates with low fines content (5%–8%, 0–40 mm), making it necessary to use an impact crusher to produce an artificial-graded aggregate for roads

Table 4.5 shows a summary of the main characteristics of the different crushers for the production of recycled aggregates.

The recycled aggregates must be stored separately in accordance to the intended used. It is important to separate the RAC aggregates from the fines, because the presence of unhydrated cement in the fines will create a risk of cracking the materials.

4.2.11 Recycling of fresh concrete waste

Recycling of fresh concrete waste is now common practice in most ready-mixed concrete plants. Such wastes may be surplus concrete or rinse water, which has been used to flush and clean mixing plants or transport vehicles.

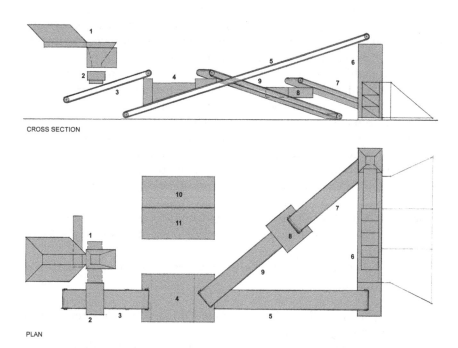

Figure 4.11 Principle design of a crushing plant. 1. Primary crusher. 2. Magnet. 3. Transport band. 4. Aquamator (water sorting). 5. Transport band. 6. Sorting plant. 7. Transport band. 8. Secondary crusher. 9. Transport band. 10. Generator. 11. Control unit. (From Lauritzen, E. K. and Jacobsen, J. B., Nedrivning og Genanvendelse af bygninger og anlægskonstruktioner, SBI Anvisning nr. 171, 1991 [in Danish].)

Figure 4.12 Setup of a typical mobile plant for crushing concrete debris to recycled aggregates. The crushing plant was established in the Beirut Harbour area to crush debris from clearing of the war-damaged Beirut Central District in 1996 (see Chapter 6, Section 6.6).

This rinse water is water used to flush out the remaining concrete from vehicles, or to separate aggregates from the fresh concrete waste by wet screening. According to Danish standard specifications, such wastewater may be used as mixing water in the production of new concrete for passive and moderate environments, provided it does not contain more than 14% suspended solids.

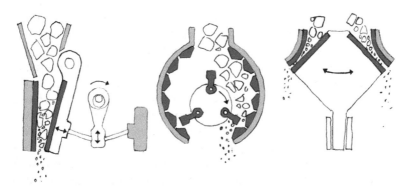

Figure 4.13 Principles of the crushers, jaw crusher (left), impact crusher (middle) and cone crusher (right). (From Lauritzen, E. K. and Jacobsen, J. B., Nedrivning og Genanvendelse af bygninger og anlægskonstruktioner, SBI Anvisning nr. 171, 1991 [in Danish].)

Table 4.5 Characteristics of different crushers to produce recycled aggregates

Property	Jaw crusher	Cone crusher	Impact crusher
Efficiency	High	Low	Medium
Production cost	Low	Medium	High
Wearing	Low	Low	Medium
Aggregate quality	Low	Medium	High
Fines content	Low	High	High
Energy consumption	Low	Medium	High

Source: Alaejos, P., S. Juan, Marta, R., Jorge, D., Roberto, V., Ignacio, Quality assurance of recycled aggregates, Progress of Recycling in the Built Environment. Final report of the RILEM Technical Committee 217-PRE, 2012. Published by Springer 2013.

4.2.12 Market for recycled concrete

In order to meet the challenges and to exploit the opportunities for recycling concrete, we must look at the specific solutions for the establishment of a sustainable and profitable market for recycling of concrete. The key elements are the following:

- Economy
- Quality of materials
- Information and education

4.2.12.1 Economy

Under the conditions of a market economy, the choice between recycled and natural materials depends on price and quality. The quality of concrete with recycled aggregates can be the same as that of concrete with natural aggregates, but recycled concrete aggregates are regarded with suspicion. Hence, recycled concrete materials will only be preferred when the price of such aggregates is considerably lower than that of the natural materials, even when the recycled aggregates meet given specifications. Vanderley (Vanderlev and Angulo 2012) states that '*Natural aggregates are simple and cheaper to process than recycled aggregates, since natural raw material is more homogeneous and has much lower contamination than recycled one. The low cost of natural aggregates has been certainly a limitation factor for a decisive factor for any material, even for sustainable construction*'.

Cost of natural materials, such as sand, gravel and aggregates depends on the production costs and the cost of transportation. The transportation depends on the availability of resources. In many building scenarios with shortages of resources, the transportation costs are high, which make recycling competitive. See Section 4.10 'Global View of the Recycling Market' and Chapter 5 'Integrated CDW Management' for further discussion on this issue and business cases exploiting opportunities for recycling on-site and reduction of transport distances.

According to Bio et al. (2011), recycled concrete aggregates in Europe can sell for 3 to 12€ per tonne, with a production cost of 2.5 to 10€ per tonne. The higher selling prices are obtained on-sites where all CDW is reclaimed and maximum sorting is achieved, where there is strong consumer demand, as well as lack of natural alternatives and supportive regulatory regimes. Crushed concrete for recycling in new concrete (RAC) can sell for the double price of RCA. In countries rich on easily accessible resources, this might not be an issue of interest. In other countries with scarce resources, the recycling of concrete waste materials might be of highly prosperous and socio-economic relevance.

The EU IRMA project (IRMA 2006) comprised a study on the economic benefit of recycling crushed concrete in various applications from back filling (low) to unbound RCA application (medium) and high-quality RAC (extreme). The results of the study are illustrated in Figure 4.14. It shows that the highest relative savings/benefits are in the medium–high interval of the application grade scale. This corresponds to the RAC of the low exposure class (see Table 4.2). It also shows that the extreme application is not favourable, because of the cost of high-quality RAC production. It should be noted that the actual shape of the curves and the location of the

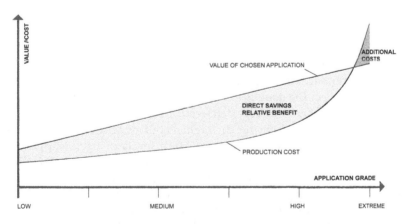

Figure 4.14 Illustration of relative benefits of secondary materials by level of application. The figure shows that the most profitable application is low quality/low exposure class RAC. The high-quality RAC of extreme application grade requires additional cost because of the higher production cost. (From Integrated Decontamination and Rehabilitation of Buildings, Structures and Materials in Urban Renewal [iRMA], European Fifth Framework Programme Energy, Environment and Sustainable Development Key Action 4: City of Tomorrow and Cultural Heritage. Contract no. EVK4-CT-220-00092. Project files and report to the EU Commission 2006.)

point of intersection between the two curves strongly depends on national legislation, practices and market conditions.

From the figure above, it is clearly illustrated that extreme application grades, such as for example direct reuse of complex building elements, eventually might carry large additional costs. The additional costs specifically associated with this segment can be ascribed to a more complex demolition/procurement, cleansing and cleaning process and the enhanced pre-conditioning of the material, as these materials are often reused in their original form, shape or 'condition'.

Not surprisingly, this type of application, together with other extreme application grades, have not gained any significant foothold in the various countries in Europe, and are possibly the least commonly used methods of reuse/recycling of CDW wastes materials in Europe. High-grade applications are also to a large extent associated with significant costs, due to costs required for preparation and quality control of the secondary raw materials. However, it should be noted that 'direct' reuse/recycling in same type of application and similar environment, that is the application of crushed concrete pavement materials as aggregates for concrete in new pavements, and so on, will reduce the associated costs significantly even for high-grade application. Low to medium grade applications are in general valuable, with costs in material preparation which rarely exceed the generated benefits. It should be noted however, that environmental aspects in certain cases could limit the application potential, or increase the cost of material preparation prior to use.

4.2.12.2 Quality of materials

The degree to which CDW is processed and recycled aggregates are used, as intended by standards and legislation, depends largely on the way authorities control and inspect the demolition processes, processing plants, the permits and the quality of the products delivered. Lack of control means a greater chance for fly-tipping, mixing different types of waste, contaminations in CDW and low-quality aggregates. High levels of useful applications of high-quality materials can be reached by paying attention to the entire process of demolishing to application by governmental organizations. The application of aggregates throughout Europe should be universal. To accomplish this, European legislation on the use of recycled aggregates and European Standards have been developed. CE-labelling of recycled aggregates or a similar kind of labelling is obliged. Although the conditions for a universal use of recycled aggregates in Europe are present, it has been noticed that in practice the use of recycled aggregates is often still controlled by national or local regulations. It is changing, but slowly. The main reason is that processing CDW and trading aggregates is not happening on an international (European) market. Due to relatively high transportation costs (compared to the costs of processing) recycled aggregates are mostly applied within a limited distance of the processing plants. Only when transportation is possible by ship is international trading of aggregates (economically) possible. Again, there are actions to be taken by governmental organizations and branch organizations. These organizations should communicate to the market, especially to the processing plants and end users of recycled aggregates that only application of CE-labelled aggregates is allowed, and therefore European standards should be followed. Furthermore, only producers of recycled aggregates, which are compliant to European standards, should be commissioned for the delivery of recycled aggregates for public works. References to CE-labels and European standards should be incorporated in the demands for quotations. Processing plants situated in areas near the borders could develop some international business. However, the European countries do not agree on the definition of recycled aggregates. Some countries still consider (some) recycled aggregates as waste. Other countries consider recycled aggregates as a product or a raw material. When recycled aggregates are considered to be waste, international transportation is not possible without an EVOA permit (international permit on the export of waste). Procedures to receive such a permit can be seen as an impediment for the trade of recycled aggregates between countries. Therefore, a clear European statement on the status of recycled aggregates, compliant to European standards could help to increase the possibilities for the use of these products.

A number of technical standards or specifications for the application of aggregates in general exist at the European level; however, currently only few standards have been developed specifically addressing the application of recycled aggregates. Some national or local standards exist, but large variations can be observed, both regarding areas covered, applications and application classes, boundary limits, and so on, from country to country and even at regional level (e.g. Belgium). Furthermore, these standards are consistently prepared in the national language, which makes a European common approach more difficult.

4.2.12.3 Environmental impact

Table 4.6 presents the typical energy consumption of production concrete. The energy consumption for cement production (82.4%) is the dominating energy consumption. The greenhouse gas emission follows the energy consumption. The typical greenhouse gases emissions for the manufacturing of cement are estimated at 0.6 tonne of CO_2 per tonne of cement, thanks to the efforts to increase the efficiency of the firing process in the cement kiln. Moreover, the use of CO_2-neutral fuels (30% of the total fuel used during the production process) and the use of alternative materials have led to substantial CO_2 savings. Experiments carried out in northern Europe have shown that the level can be lowered to 0.35 tonne of CO_2 per tonne of cement by implementing even further these measures (Bio et al. 2011).

Table 4.6 Energy consumption associated with each
concrete constituent

Constituent	Energy (MJ/kg concrete)	Energy (% distribution)
Coarse aggregate	0.028	3.1
Fine aggregate	0.028	3.1
Portland cement	0.735	82.4
Water	0.000	0.0
Manufacturing	0.102	11.4
Total	0.893	100

Source: Bio Intelligence Service, Acadis, Institute for European Environmental Policy. Service contract on management of construction and demolition waste – SR1, Final Report Task 2, February 2011.

Table 4.7 presents results of a case study of energy consumption and emissions to air from concrete made by natural aggregate (NAC) and recycled aggregates (RAC). The results show that cement production is the largest contributor to all impact categories, for both NAC and RAC. It causes approximately 81% of the total energy use and 92% of the global warming. The main reason for this is substantial CO_2 emission during the calcination process in clinker production, and the use of fossil fuels. It should be noted that the energy consumption and emissions of concrete production are equal for NAC and RAC concrete production. According to the explanation of the case study, the ratio of transport distances of RCA to NA was assumed to be 15 km:100 km, which means that the recycling plant must be located much closer to the concrete plant than the place of NA extraction, if environmental benefits from recycling are to be expected (Pacheco-Torgal et al. 2013).

As previously mentioned, no general European environmental specification, standardization of regulation exists or is available for neither the use of primary nor secondary or recycled aggregates. In general practice, local or national standards or specifications are applied, when in existence. Large variations exist, however, between the different countries in Europe, and sometimes differences are even observed between regions within the same country (e.g. Belgium and Germany). The variations observed span from differences in availability, severity, field of application and type of specification to basic requirements on boundary limits. Common to most of the national or regional environmental regulations, however, is that the main focus is on the actual content of a given substance in the individual products rather than on the actual release of substances.

The variations, in combination with the use of local languages, inhibit the potential for import/export of materials and knowledge across borders as well as the potential for application of materials or recycling procedures from

Table 4.7 Inventory table of 1 m³ concrete natural aggregates and recycled aggregates

	Cement (kg)		Aggregate (kg)			Concrete (1 m³)		Transport (1 m³)		Total (1 m³)	
			NAC	RAC							
	NAC	RAC	NA	NA	RCA	NAC	RAC	NAC	RAC	NAC	RAC
	354	365	1,764	576	1,071						
Energy (MJ)											
Coal	1,193	1,230								1,193	1,230
Diesel	8.63	8.89	26.07	8.51	56.90			269.30	253.06	304.0	327.37
Natural gas	29.45	30.36								29.45	30.36
Electricity	179.72	185.30				20.07	20.07			199.8	205.37
Emissions to air (kg)											
CO	1.488	1.534	0.006	0.002	0.012	0.001	0.001	0.066	0.063	1.561	1.612
NOₓ	0.807	0.832	0.027	0.009	0.044	0.013	0.013	0.176	0.158	1.023	1.056
SOₓ	1.291	1.331	0.009	0.003	0.018	0.099	0.099	0.076	0.071	1.475	1.521
CH₄	0.354	0.366	0.002	0.001	0.005	0.001	0.001	0.022	0.020	0.379	0.392
CO₂	304.9	314.3	2.430	0.739	4.506	5.698	5.698	19.40	18.20	332.4	343.5
Particles	0.252	0.259	0.003	0.001	0.007	0.012	0.012	0.042	0.027	0.308	0.306

Source: Pacheco-Torgal F., T., W.W.Y., Labincha, J.A., Ding, Y. and de Brito, J., Handbook of recycled Concrete and Demolition Waste, Woodhead Publishing 2013.

NAC: Natural aggregate concrete, RAC: Recycled aggregate concrete

NA: Natural aggregate, RCA: Recycled concrete aggregate

other countries, and call for common European guidelines or specifications. However, the degree of variation observed might prove the standardization process very difficult.

4.3 RECYCLING OF MASONRY

Bricks, masonry and tiles are common terms for many different products. Masonry is a structural composition of bricks and mortar, while bricks and tiles are individual building materials. In this book, masonry is also used as the general term for all kinds and compositions of bricks and tiles in the construction industry. Bricks and tiles are produced by burning a mixture of clay, sand and ancillary materials such as sawdust, lime, barium and manganese. Thus, the raw materials composition of masonry is such that recycling of the clean fractions in CDW is unproblematic from an environmental point of view. Most brick and tile products are potentially recyclable. However, the amounts of bricks and tiles waste, the associated recycling and reuse rates are not available at the EU-27 level due to the absence of a systematic European reporting system (Bio et al. 2011) (Table 4.8).

Bricks, masonry and tiles can be recycled in different ways:

- Reuse of whole bricks, roofing tiles and slates
- Crushing of bricks, masonry and tiles to produce secondary raw materials for use as:
  - Aggregate in new concrete
  - Unbound road-base materials
  - Other applications

Table 4.9 presents options for recycling bricks, masonry and tiles.

4.3.1 Reuse of whole bricks

Manual cleaning and reuse of whole bricks has been common practice through the ages. Today, manual cleaning of bricks is neither economically viable nor environmentally sound. Another important issue is that only bricks from masonry constructed with lime mortar can be separated

Table 4.8 Hierarchy of recycling bricks, masonry and tiles

Ranking	Reuse, recycling, recovering
1	Reuse of bricks, masonry elements and tiles
2	Recycling of crushed bricks and tiles in new products, bound or unbound
3	Recycling of crushed bricks and tiles in road base
4	Recovering as backfill, landscaping, walls, etc.

Table 4.9 Options for recycling bricks, masonry and tiles

Rank	Possible use	Comments
I	Reuse of bricks	Cleaned bricks are suitable for new façades. Only bricks from old masonry built with chalk mortar, or mortar with low cement content, can separate. Because of the standard size, reuse of bricks is flexible
I	Reuse of roofing tiles	Old tiles are used for new roofs or for repairing of old roofs. Because of individual type and measures, the opportunities for reuse are limited
2	Aggregate for production of new concrete	Crushed bricks are used as coarse aggregate (4/32 mm) for concrete in passive environment • The fraction 0/4 mm cannot be used for production of concrete • The materials must comply with the RILEM recommendation (RILEM 1994)
2	Cement bound base courses	Little documentation is available and only in the form of results of laboratory experiments
3	Unbound base courses	• Recommended for use in roads classified for 0 to light vehicle traffic • Not recommended for use in roads classified for medium to heavy traffic
3	Fill around conduits and in trenches	• Properties of the fraction 0/4 mm have been studied • The results show that the percolation properties of the material comply with common requirements. The load bearing properties of the material have not been studied
3	Support for flagstones	In pedestrian areas only
3	Moisture barrier courses	Capillary rise of crushed bricks is too high
3	Filtering material around drains	Percolation properties need further studied
3	Green roofs	Crushed tiles in a bio compound are used for roofs with small vegetation, green roofs
3	Other purposes	Cat gravel, tennis court, etc.
4	Backfill	Backfill, building foundations, landscaping, etc. In addition to the direct use of crushed bricks as fill, soil can be mixed with bricks in order to lower the water content of the soil and to improve its compaction properties

and cleaned; it is not possible to separate brickwork build with cement mortar. In the early 1990s, the Danish Environmental Protection Agency (DEPA) funded several research and development projects of equipment for mechanical cleansing of bricks. Unfortunately, the machines did not function properly; it was very difficult to clean the bricks without damaging

Figure 4.15 Example of recycling bricks from demolition of buildings at the Bispebjerg Hospital, Copenhagen, for construction of façades of new buildings, Katrinedals School. The figures show the processes of recycling bricks from demolition (top) to transformation including cleaning and preparing the bricks for reuse at the Danish company. Old bricks (middle) and reuse of the bricks for construction of new school building. (From Copenhagen, Municipality of Copenhagen, Genbrug af mursten. Erfaringer fra nedrivning af bygning 16 på Bispebjerg Hospital og genbrug af mursten til nybyggeri på Katrinedal Skole, Vanløse, 2017 [in Danish].)

them. After the initial developments and tests, the Danish company Old Bricks (*Gamle Mursten*)[5] invented in 2004 a cleaning concept based on a flexible abrasive band shown in Figure 4.15.

In 1995, reuse of bricks was technically accepted but not economically advantageous. Today, reuse of bricks is commonly accepted and preferred in many projects, promoted by public trends of recycling and the circular economy. Figure 4.15 shows practical example of reuse of bricks.

Compared to roof tiles and floor tiles, bricks in walls, which not have been exposed to extreme loads, are everlasting. Therefore, old bricks do not lose their value; on the contrary old stones might often have an architectural preference compared to new stones. Based on practical experiments, test and documentation, the reuse of old bricks is now generally accepted in Denmark as a competitive façade material to new bricks. This makes room for reused

---

[5] www.gamlemursten.dk

bricks in the professional market for building and construction materials. It should be noted that the buildings at present time are usually constructed with bearing structures and façade elements made with concrete, and that the façades made with bricks are not bearing structures. Therefore, the strength of the reused bricks is not a crucial issue. An example of circular economy with the demolition of masonry and reuse of whole bricks is presented in Chapter 5, Section 5.3. The development and experiences with reused bricks are described in the publications of the Municipality of Copenhagen (Copenhagen 2016; Copenhagen 2017 [in Danish]).

The Danish Technical University has made a Life Cycle Analysis (LCA) (Møller et al. 2013) of the reuse of whole bricks compared to crushing and recycling bricks and masonry. It has found that recovery of masonry waste with regard to reuse of bricks has environmental advantages. Presupposed that 64.5% of masonry waste is reusable bricks in new façades, 103.4 kg CO_2-eqv. per ton of masonry waste is saved compared to recycling crushed masonry; 64.5% bricks of one ton masonry is equal to approximately 269 stones, each 2,4 kg, per ton masonry. In this specific case, it is estimated that we will get a CO_2 saving of approximately 0.4 kg CO_2 per brick. The CO_2 saving by reusing bricks for other purposes than façades is lower, approximately 0.2 kg CO_2 per brick. The LCA report concludes that substituting façade bricks with reused bricks is a more environmentally advantageous solution than recycling crushed masonry (Møller et al. 2013).

Looking at the economy of reuse of bricks, the cost of sorting and cleaning bricks by hand is not competive with production of new bricks in industrial regions. Recycling of bricks by use of the technology invented by *Old Bricks* matches the price of new bricks of medium quality.

Mixtures have been produced of various clays with different contents of calcium and iron, which cover most compositions found in old recycled bricks in Denmark. All clay mixtures have been burnt to temperatures between 900°C and 1100°C. The body colours are reproduced in a series of illustrated plates, which make it possible to evaluate the quality of most common bricks produced in the country over the centuries.

It has been attempted to re-burn whole bricks with masonry mortar attached. It was found that all lime mortars and cement-lime mortars could be completely removed from the bricks by re-burning. However, the re-burning of whole bricks with strong cement mortar attached gave rise to severe cracking of the bricks. High-energy consumption was another disadvantage of the re-burning process. Further development of the process has been discontinued.

4.3.2 Recycling of whole roofing tiles and slates

As is the case for the reuse of whole bricks, it has always been obvious to reuse whole roofing tiles and slates. This has proved to be feasible from a technical as well as an economic point of view. However, such reuse does require:

- Selective demolition with careful removal of the tiles and slates from buildings
- Manual cleaning of tiles and slates assisted by high-pressure water flushing, which unfortunately involves health and safety hazards

The major problem and barrier to the reuse of whole roofing tiles and slates is the huge variation of types and products of roofing tiles and slates. While bricks vary with respect to colour and quality and a limited number of standard sizes, tiles and slates are more or less individual. This requires a lot of sorting work, areas for storage of the many types of items together with a great amount of administrative work. Therefore, there is only room for reused roofing tiles and slates on the private market – not on the professional market of building and construction products.

4.3.3 Crushing of bricks, masonry and roofing tiles to secondary raw materials

The different recycling options promoted by the European Tiles and Bricks Association are described below in the Bio report to the European Commission (Bio et al. 2011):

- Crushed clay bricks, tiles and other masonry can be used as aggregate in concrete. The crushed material replaces other raw materials such as sand. This is commonly practiced in Austria, Denmark, Switzerland and especially The Netherlands
- Crushed clay bricks, roof tiles and other masonry can be used on larger road building projects, especially as unbound base material. It is used to build roads in countries such as Germany, Denmark, The Netherlands, Switzerland and the United Kingdom. In Germany, the maximum brick content for such use is 30%, due to quality requirements for frost attacks and impact resistance. The material replaces natural materials, such as sand and gravel, which are normally used in large amounts for this purpose
- Crushed clay bricks and other masonry can also be used to level and fill pipe trenches. The fine crushed material will replace natural materials such as sand
- Tennis sand produced by crushing red bricks and roof tiles. The fine surface layer is laid over courser-grained layers that can comprise crushed clay brick matter. The process is most efficient when it occurs at brick or tile factories where there is an abundance of scrap material

- Crushed bricks and tiles can also be used as plant substrates. The material may be mixed with composted organic materials and is especially suited for *green roofs*: the porosity of the material allowing retaining water plants can rely on during dry periods
- Backfill and stabilize minor roads, especially in wet areas such as woods and fields. The practice is common in countries that lack adequate stone supplies such as Denmark. The material is generally used uncrushed

4.3.4 Aggregate for production of new concrete

Referring to RILEM Report No. 6, Schulz and Hendriks have provided a comprehensive report on recycling of masonry rubble (Schulz and Hendricks 1992) focusing on crushed masonry as aggregates in new concrete. This research has contributed to the RILEM *Specifications for Concrete with Recycled Aggregates* (RILEM 1994) addressing crushed concrete and masonry to be used as aggregates in new concrete. EU and national standards for raw materials permit crushed masonry to substitute for natural aggregates in new concrete.

In Denmark in 1989, a full-scale test was carried out for the purpose of demonstrating the use of crushed masonry as aggregate for the production of new precast concrete elements for an industrial building (see Figure 4.16). The masonry was delivered from the selective demolition of a 100-year-old factory. It was crushed and screened to provide a 4/16 mm material. Crushing gave rise to a large wastage rate due to the generation of fine material below 4 mm. The results of the project showed that it is possible to use crushed masonry in industrial production of new concrete elements. However, if the compressive

Figure 4.16 Industrial building with facades of elements with crushed masonry aggregates, Copenhagen 1989.

strength of the masonry particles is lower than the required strength of the concrete, the mortar must be correspondingly stronger. At the time there were no reliable methods available for the mix design of concrete with crushed bricks as aggregate. In that respect, the problems are similar to those of lightweight aggregate concrete. The high water absorption of crushed masonry also gives rise to problems, because the methods normally used for determining water absorption of natural aggregates are not applicable to crushed masonry.

The results of the practical tests are inconclusive for what concerns economic viability of using crushed masonry as aggregate in the production of new concrete. Transport costs of recycled versus natural materials are important parameters, and it would be necessary to ensure a reliable and continuous production of certified crushed masonry aggregate in order for the use of such aggregate to become viable.

A demonstration project in Denmark on 3D printing a minor house has shown opportunities using crushed masonry below 4 mm in concrete for 3D printing (see Chapter 7, Section 7.2).

4.3.5 Unbound aggregates for road construction

It is an obvious idea to use crushed bricks and tiles for road construction purposes and on paved sites. However, results of research projects have shown that in Denmark crushed bricks are of such variable a quality that unbound crushed bricks should only be used as common unbound base course material on minor roads submitted to light traffic. Based on practical trials, the Danish Road Directorate has issued instructions and standard specifications for the use of crushed bricks in unbound base courses. When used in cement-bound base courses, the quality of crushed brick rubble is usually equivalent to that of natural gravel. However, use of very weak crushed bricks in cement-bound base courses has given rise to a large drying shrinkage and formation of reflection cracks in the overlying asphalt surfacing.

Use of pure crushed bricks and tiles as road construction material has never been popular in Denmark. However, the above-mentioned specifications allow a maximum content of 45% concrete and mortar in the crushed brick fraction. In actual practice, a mixture of crushed masonry and concrete is being produced at recycling plants and used on lightly trafficked roads, rather than pure crushed bricks and tiles. The blend of bricks and concrete has come into common use because mixtures of bricks, masonry, tiles and concrete rubble frequently are received at recycling centres rather than the pure materials fractions. The mixed material has high strength. It can be used under very wet conditions, and it offers good support for temporary access roads. Overall, it has better properties than pure crushed bricks and tiles. Therefore, it has been decided at recycling centres to produce the much-demanded mixed material rather than attempting to separate the waste into pure fractions.

4.3.6 Other applications for recycled masonry

Studies have shown that the 0–4 mm fraction of crushed bricks can be used as support for flagstone pavements and as filtering material around drains. Crushed bricks are also useful as fill in cable conduits and trenches. However, because of a high porosity and a resulting high capillary rise of the material, crushed brick rubble should not be used as sub-base material in road construction or in other types of moisture barrier courses. It has been shown that crushed bricks can be used as aggregate in new concrete up to a compressive design strength of maximum 20 MPa. Table 4.9 summarizes the possible practical applications of crushed bricks and tiles.

4.3.7 Market for recycled bricks, tiles and masonry

The market for recycled bricks exists. There is a demand for façades of old bricks in new buildings, not as a bearing structure but only as the façade cover. Old bricks are also used indoor as a part of the interior decoration, especially in café's and restaurants. Furthermore, old bricks are used for repair and reconstruction work of old houses. Presupposing appropriate logistic and supply of bricks from demolition works and industrial cleansing, reused bricks can be marketed for a competitive price of 1–2 EUR per brick, equal to the price of a medium quality new brick. Hand cleaned bricks are much more expensive, 3–4 EUR per brick. The reused tiles market is more difficult, because the big variation of tiles and individual demand of tiles.

4.4 RECYCLING OF ASPHALT

Asphalt is the common term for mixtures of a binder and a granular material of mineral origin, usually crushed natural stone. In earlier times, only tar was used as binder. Now bitumen, or a mixture of bitumen and water, is used. Fatty amines are added in order to improve the adhesive properties of bitumen. Asphalt is used in road construction as a wearing or base course material. Waste asphalt is generated when roads are reconstructed, rerouted or repaved.

It is assumed that bitumen, which is an oil distillate, may contain residues of oil. Nevertheless, the Danish Environmental Protection Agency considers that the use of crushed asphalt for road construction purposes does not constitute a hazard to the environment. It is foreseen and recommended by the authorities that recycling of asphalt should primarily take place in the production of new asphalt, either in mobile plants on-site or in stationary plants. However, there will always be some remaining material, which for practical reasons cannot be used in new asphalt pavements.

In principle, broken asphalt can also be used as sub-base material and as fill. However, it should be kept in mind that compaction of demolished asphalt is difficult, and that permanent deformations under static load are larger for crushed asphalt than for natural stabilized gravel. However, practical experience has shown that a mixture of 50% crushed asphalt and 50% crushed concrete provides an excellent unbound road base material. The lubricating properties of asphalt reduces the effort required to compact the material by rolling. The mixed granular material is stronger than compacted crushed concrete and it supports static loads much better than pure crushed asphalt.

Hendriks et al. (2000) and Bio et al. (2011) present recycling of asphalt as follows. Three major types of asphalt are distinguished depending on the production temperature: hot, warm and cold mix asphalt. The primary use of asphalt is in road infrastructure construction and in airports for runways and therefore is referred to as asphalt pavement. Over 90% of the total road network in Europe is made of asphalt. Production takes place in a fixed or mobile mixing plant with two main processes namely in batch plants and in continuous mixing or drum mixers. Recycling means adding the reclaimed asphalt to new asphalt mixes, with the aggregates and the old bitumen performing the same function as in their original application. Therefore, reclaimed asphalt replaces virgin aggregates and part of the binder.

The recycling processes can be divided into two major methods: hot or cold mix recycling techniques. These can be further sub-divided into stationary plant or in situ recycling. To achieve the highest levels of recycling it is necessary to either confirm the lack of variability in the feedstock or to have precise data on its range of properties. The requirements for reclaimed asphalt are formulated in the European Standard *EN 13108–8 Reclaimed Asphalt*. The difference between the cold and hot recycling methods only relies on the process to heat reclaimed asphalt pavement (RAP). In the case of the 'hot method' in a hot mix recycling stationary plant, the RAP taken from demolished or renovated roads in general is transported to the asphalt plant. After being crushed and sieved (if needed), the RAP is directly preheated, which requires an extra dryer (a dryer being already used to heat the virgin aggregates that are then incorporated into the asphalt mix). Incorporation rates for the hot method are typically 30%–80%, the upper limit being determined by the quality requirements of the mix specification in relation to the properties of the old asphalt. A variation of the hot recycling is feeding the RAP into the same dryer as for virgin aggregates (this method is called a *recycling ring*). The heating of the RAP takes place behind the flame, ensuring that it is not overheated. This method allows up to a 35% incorporation rate. It is possible to use a recycling ring in combination with a rotary drum dryer, thus achieving incorporation rates up to 50%. The cold method is hot mix recycling in a stationary plant. As for the hot method, reclaimed asphalt is crushed and screened (if needed) so as to produce a consistent feedstock. The difference between hot and cold methods in hot mix recycling relies on the process to heat RAP: in the first case, a specific dryer is needed while in the other case reclaimed asphalt is heated through the contact with heated virgin aggregates. In both cases, the reclaimed asphalt is used to produce new hot mix asphalt. The appropriate amount of new bitumen is then added in the mixing unit according to the desired end properties. Cold methods in hot mix recycling achieve incorporation rates between 10% and 40%, depending on the RAP moisture content and quality, the type of the plant's vapour extraction system and the technical process limitations regarding maximum permitted temperatures. Modifications to the plant are needed if quantities of more than approximately 10% of old asphalt are to be added to the mixing process. Though tar is not used anymore in road construction, amounts of asphalt containing tar can still be reclaimed when renovating old roads. In this case, the waste is considered hazardous and the hot recycling is not allowed. In some countries, however, it is allowed to rely on cold techniques with or without binders.

4.5 RECYCLING OF WOOD

Waste wood is only partly recyclable. Yet many wood products can be reused in their existing dimensions and shapes. Cleaned and nail-free timber and boards can be reused in new construction, and uncontaminated wood can be shredded for use in horticulture and agriculture, or in the production of parquet boards, veneer and panel boards. However, the large quantities of painted and impregnated wood arising from demolition are considered polluted material and must be treated as chemical or hazardous waste due to risks deriving from the leachates.

4.5.1 High quality of old wood

The quality of most old timber and wood products has proved to be good and is able to compete with that of new materials. Old building materials often consist of high-quality wood, and often use the best heartwood for more exposed parts such as windows. Old timber is often of dimensions needed in the renovation of old houses, and often displays a patina that is required to restore or create a genuine environment. In addition, used timber does not require nearly as much proofing chemical as new wood, yet a further bonus for the environment (Table 4.10).

The *Recycled House in Odense*, which is described in Section 4.9, is an example of the reuse of wood in construction. The overall problem and barrier to the highest level of recycling is the need for manual resources to recover the reusable timber, comprising extraction of nails, clean cutting, cleansing, and so on. Furthermore, there is a market problem, because reuse of wood is normally only attractive to private costumers, unless we are speaking of a major stock of timber and planned reuse before the start of

Table 4.10 Hierarchy of recycling wood

Ranking	Reuse, recycling, recovering
1	Reuse of timber in new construction or repair of existing structures
2	Reused timber sawn into new products, construction timber and floorboards
3	Reuse of processed products, wooden rafters, skirting boards, architraves, floorboards, windows, doors, kitchen and cloakroom, cupboards, doors and windowsills
4	Recovering by shredding of wood for agriculture and gardening purposes
5	Energy recovery by incineration

the demolition process. It is often a problem to find the right quantity and the right quality of old timber at the right time and the right place.

Old timber is often harder than new timber (see Figure 4.17), therefore other types of machines are required for sawing. In addition, machines are frequently damaged by nails left in the wood. Competent personnel is needed to saw up timber into smaller dimensions.

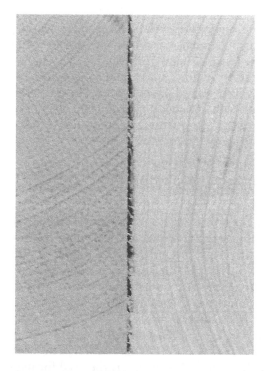

Figure 4.17 Photos showing old wood (left) and new wood (right). The distance between the annual growth rings in the new wood are considerably larger than the distance between the growth rings in the old wood, which means that the new wood has had a shorter period of growth and therefore has less strength than the old wood. (Lauritzen and Hansen 1997.)

Modern building processes often comprise an assembly procedure with a tight time schedule where the fitters and building workers neither have the experience nor the time to undertake a systematic evaluation of different wood products. This means that there are no resources available to evaluate, for example, which way a beam shall be turned, or if it requires extra support. Today's building process requires timber in standard dimensions and of uniform quality, which further limits opportunities for reuse of old timber. In each specific case, the question must be asked whether it is economical to saw the timber into new products, or whether it should be used in its existing dimensions. On many projects, the re-sawing of used timber has proved to be very expensive. Consequently, due to traditions and psychological barriers, the general attitude towards recycling inhibits wider utilization of old wood.

4.5.2 Problems with reuse of poor-quality wood and wood attacked by rot and fungi

It has been ascertained that much timber from demolition works is suitable for reuse. However, there are problems related to the reuse; the major one being that of rot and fungi attack. The successful reuse of timber therefore depends on the correct recognition and solution of these problems. For that reason, it is necessary to train workers involved in the selective demolition process to enable them to recognize rot and fungi attack during demolition, and to sort out contaminated timber at the source.

Apart from biological attack, timber for reuse should not have been exposed to structural damage caused by impacts or maltreatment. Careful attention must be paid to the presence of shrinkage cracks and holes, and nails must be detected and removed if the wood is to be re-sawn. Earlier exposure of wood to high temperatures or chemicals may also preclude any reuse. When timber is to be used in construction works, it is necessary to investigate and classify its quality. Structures built with reused timber must meet the same criteria for what concerns strength in structures using new timber.

The properties of building materials from old structures can differ from new materials. Therefore, existing selection criteria for new materials cannot be applied to recycled materials without further consideration. This is most critical for what concerns selection methods that only give an indirect indication of the material's quality, such as ultrasonic tests. Therefore, there is a need for revised sorting criteria and classification methods, including general guidelines for the reuse of old wood.

4.5.3 Problems with recycling of wood contaminated with chemicals

After selective demolition, wood is sorted into a non-contaminated fraction and a contaminated fraction. Reuse of non-contaminated wood as floorboards, and reuse of timber or windows and doors from older structures seldom give rise to serious problems. Neither does disposal by incineration or recycling of chemically non-contaminated wood in the production of particle- or fibreboard. However, much demolition wood contains non-wood contaminants such as concrete, plaster and asbestos, as well as organic soil, adhesives, laminates, paints, preservatives and fire retardants, enriched by compounds not used nowadays because of their harmful properties such as arsenic, lead, mercury, chlorinated aromatics, and so on. Usually wood preservatives are the most disturbing impurities.

In central Europe, an increasing amount of waste wood is being used as raw material, especially for the production of particle- and fibreboards. Efficient sorting and purification procedures are essential to avoid hazardous contamination of the new panels. In Denmark, contaminated wood is disposed of by incineration or at a controlled site.

4.5.4 Recycling of windows

In connection with a demonstration project within the Danish recycling effort, various ways of recycling wooden windows were studied. Such reuse proved to be more difficult than initially anticipated.

It was concluded that none of the old windows used in the new buildings were rainproof according to modern standards, even after improvements were made. This is because the permeability of wood increases with age due to formation of seasoning cracks. Only new windows, which were manufactured from recycled wood, could live up to modern requirements and only so when wood of carpenter quality was used – timber quality did not suffice.

It may be concluded that old windows cannot at present be reused in new buildings in Denmark because they do not live up to stringent modern standards. Nevertheless, it was found that old windows when renovated and provided with double glass often are of very good quality. Such windows are frequently used in the rehabilitation of old buildings in order to preserve the original architectonic appearance of the buildings.

4.6 RECYCLING OF METALS

There has always been a tradition for the recycling of metals. Thus, no particular public effort is considered necessary in order to promote such recycling. Highly processed non-ferrous metals, in particular aluminium and copper, and non-reusable ferrous metals like steel reinforcement are sold for reprocessing. Reusable ferrous metals such as whole steel profiles may be used directly. Corrugated iron is sometimes reused as permanent formwork for concrete. In demolition of industrial plants, offshore installations, towers and masts recycling of metals is very important with respect to the economy (Table 4.11).

Table 4.11 Hierarchy of recycling metals

Ranking	Reuse, recycling, recovering
1	Reuse of construction steel, and steel structures in new constructions or repair of existing structures
1	Reuse of heating and sanitation steel and metal, tubes, tabs, and other kinds of installations
2	Reuse of construction steel, such as I, U, H profile steel for supporting structures and temporary constructions
3	Recovering of scrap iron and metals, melting and production of new materials
3	Recovering of metals from electronic waste

Generally, recycling of metals comprises:

- Reuse of construction steel, roofing plates, etc., and reuse of instruments, machines and other metal structures
- Recovery of metals, scrapping steel, aluminium, stainless steel, copper, etc.

Reuse of construction steel, such as major dimensions and common types of H-, I- and U-profiles is normal. Reuse of instruments, machines from industrial plants depends on the actual situation and demand on the market. Demolition of production plants might in some case raise opportunities for selling specific machines or installations to other product plants with special needs. Typical products such as cranes, lifts, electric motors might be sold for reuse. In developing counties, it is common to crush concrete structures to recover reinforcement bars for reuse in new structures.

The price of scrap steel and metals depends on the world market, which fluctuate widely depending on the political and economic situation. From time to time, change of prices of up to 50% are not exceptional. Therefore, the price of demolition of industrial plants and offshore structures with a high content of steel and metals depends on the scrap price and the actual risk of decline/increase of prices during the progress of the demolition work. Therefore, it is seen that demolition contractors focus all resources on the recovery of scrap metal as soon as possible after overtaking the worksite, or they might prefer to store the scrap metal and wait for better prices before selling the scrap metal.

4.7 RECYCLING OF OTHER MATERIALS

4.7.1 Gypsum

There are two types of gypsum: natural gypsum (which is directly extracted) and synthetic gypsum. Gypsum is a rock-like mineral commonly found in the earth's crust and produced from open-cast or underground mines. In Europe, the principal gypsum deposits are located in France, Germany, Italy, Poland, Russia, Spain, the United Kingdom and Ukraine. Gypsum is generally screened to remove 'fines' (mainly mudstones), then crushed and finely ground. Gypsum is used mainly in the manufacture of non load-bearing building elements for setting the ceiling on and dividing the interior space. The gypsum industry is therefore principally driven by the construction activity and the demand for new and refurbished housing. Gypsum-based applications range from complex high-tech systems to easy to install products (Bio et al. 2011).

According to the BIO report (Bio et al. 2011), the collected plasterboard stream has to undergo several steps in the recycling process. First, paper layers of the plasterboards are removed as much as possible, then gypsum is crushed into powder and eventually this powder is sent back to plasterboard manufacturers so that they can make new plasterboards from it. The gypsum powder is estimated to represent 94% of the total plasterboard waste collected. The remaining 6% refer to paper and cardboard (and the related contaminants) composing plasterboards and can be reused in various ways such as composting (as very little gypsum is left on the paper) or heat generation. There is always a residual paper fraction that remains in the powder and which hinders the improvement of the introduction rates of recycled powder into the processes that are currently in place. The associated risks are the damage of the manufacture machinery and an effect on the acoustic or thermal quality of the final product.

According to Eurogypsum, between 5% and 10% of gypsum powder resulting from construction plasterboard waste is re-integrated in the closed-loop system. This figure is a European average and huge differences exist between member states. Indeed, recycling practices exist in Denmark, Germany and other northern European countries while recycling is limited in Greece and Spain, or is not applied at all in eastern European countries. In some EU member states, where comprehensive gypsum recycling schemes have been established (e.g. Denmark), overall recycling rates of 65% can be achieved. This discrepancy is due to specific national legislation, to mentality aspects and especially to the existence of a market that would encourage the economic activity associated to recycling.

4.7.2 Glass

According to the Thematic Strategy on the Prevention and Recycling of Waste (EC 2011), glass represents about 0.66% of the annual production of CDW in the EU – approximately 2.6 million t (based on total CDW 400 mill. t). Glass in buildings can be reused, recovered and recycled for glass production or production of insulation materials. However, most

CDW glass is very often crushed together with the other building materials and put into landfills. Recovery of glass for production of low-quality glass or insulation materials requires that the glass be carefully cleaned, because glass production plants are very sensitive to mineral impurities. In Denmark, the demolition contractors get money for delivery of clean CDW glass to selected glass factories.

Glazing in buildings comes in many forms depending on the building type, the frame materials used, the installation period, locations (construction code), and so forth. In general, recovered building glass can be grouped into three categories:

- Glazing from large tertiary building (façades)
- Glazing from residential collective buildings and individual house windows
- Glass used for interior applications (balustrades, glass walls, mirrors, etc.)

The broad range of buildings and glass types, and the fact that glass is usually part of a framed window and not a 'stand-alone' product makes a collection scheme complex. All these aspects must be looked at when considering collecting, sorting, treating and ultimately recycling glass.

4.7.3 Mineral wool

Mineral wool, insulation material made of stone and glass, makes up less than 1% of the total amount of CDW. In Denmark, recycling of mineral wool from demolition, rehabilitation and construction deals with glass wool and products from Rockwool. Rockwool has established a take-back arrangement in cooperation with the municipal waste management systems. It is important that the mineral wool be separated from CDW and recycled or disposed.

4.7.4 Plastics

The number of different plastics, the varying recycling processes and the need for the waste material to be clean make it difficult to recycle plastics despite their ever-increasing use. A labelling system on plastics could aid in its recycling. Many demolition yards sell fittings and other miscellaneous items found on a demolition site. This may include heritage and modern times sinks, kitchens, ovens, taps, power points, fireplaces, shower screens, hot water units, wooden rafters, skirting boards, architraves, floorboards, windows, doors, kitchen and cloakroom, cupboards, doors and windowsills. Such items are frequently bought by do-it-yourself and small-scale renovators. Producers of other building products such as plate glass and gypsum boards are gradually accepting the idea of taking back their own products for reprocessing. But much remains to be done before this becomes general practice.

4.8 TECHNICAL SPECIFICATIONS AND CERTIFICATION OF PRODUCTS

Marketing of recycled products requires technical specifications and documentation of the recycled products. The demand for cheap and recycled products on the private market might lower the requirements and the technical specifications. On the professional market for building and construction products, the recycled product should fulfil the specification of the normal or primary products, and it is a prime requisite that the recycled products are clean. An important aspect to the European recycling effort has been the development of a set of norms for recycled materials such as *EN 206-2013 Concrete-Specification, Performance, Production and Conformity,* and *EN 12629 Aggregates for Concrete* for use of crushed recycled concrete as aggregate in new concrete for passive and moderate environments. Similarly, standard/guidelines have been developed for use of broken asphalt and crushed bricks as unbound road base materials. With respect to the waste characteristic of the CDW and the risk of pollution of the materials, it is very important to develop and implement EOW criteria on national, EU or federal bases.

A basic condition for recycling CDW is that the materials are clean and that they do not pollute the environment. Documentation of the sources of the materials and purity is essential to gain the confidence of the market and the building owners.

Generally, the requirement of CE marking of building products, according to the *European Construction Products Regulation* (EU Regulation 2011), is a barrier to marketing recycled products. The problem is that the original manufacturer of the product is unknown, or the manufacturer is not able to guarantee the technical quality of the product. Therefore, in case of recycling materials without reference to standards as the above-mentioned EN specifications, the recycled products must be subjected to specific approval tests in order to obtain legal acceptance of the products, such as reuse of bricks, doors, windows, concrete elements, construction timber, and so on. The introduction of special marking and testing of recycled products is under discussion in EU member states. There is a need for revised sorting criteria and classification methods including guidelines for the reuse of wood. There is also a need for specifications covering the use of crushed recycled concrete as aggregate for new concrete for severe environments for instance in pavements and as unbound base course material and sub-bases. It is recommended that such norms and specifications should be prepared on a national basis.

4.9 RECYCLED BUILDINGS, EXAMPLES OF FULL-SCALE RECYCLING

In spite of the fact that much research and development has taken place, the implantation and progress today within the CDW recycling industry in Europe, and probably all over the world, has been dominated by basic recycling of unbound materials and the sporadic reuse of building products. It is evident that the results of research and development for instance on RAC, and the development of standards for it, have not resulted in a general acceptance of construction with RAC. Recycled products are not generally accepted and placed on the public market for building products and resources. Building owners and their consulting architects and engineers are sceptic with respect to the quality and the purity of these products.

In order to bridge the gap between the academic diaspora and the building and construction sector in the field, and to create understanding and confidence among the various stakeholders, full-scale successful demonstration projects are very important. In this section, selected demonstration projects are presented for general inspiration of all stakeholders of the building and construction sector.

In the decade from 1985 to 1996, the Danish Environment Protection Agency (DEPA) implemented action plans on recycling of CDW. After completion of a number of minor projects within the Danish recycling effort, successfully demonstrating various demolition and recycling technologies, but which all had fairly limited objectives, DEPA decided to carry out larger integrated demonstration projects with broader perspectives.

By demonstration projects are understood projects of practical significance with a high degree of visibility for the general public. The objective of the demonstration projects is to prove the readiness of technology development and introduction of the results into the market. The following types of demonstration projects were carried out:

- Reuse of crushed concrete and masonry as base course and sub-base material in road construction
- Reuse of crushed concrete as aggregate in new concrete foundations
- Reuse of crushed masonry as aggregates in precast concrete elements in a new building

Starting in 1990, a full-scale pilot project was carried out for the purpose of studying the behaviour of crushed concrete as base course and sub-base material in a road, carrying light traffic in a residential district of a provincial town. It is the main conclusion of the project that the quality of recycled concrete depends on the quality of the original concrete from which it is derived. However, crushed concrete of normal structural quality will always provide road base and sub-base materials of a quality equivalent to natural gravel both in the construction and service phase of a road. This is the case even if neither the sorting out of soil, wood and plastics from the concrete, nor the crushing of the concrete is strictly controlled during production of the recycled granular material.

In 1989, two old concrete railway bridges were demolished and the concrete was crushed and screened into a 4/16 mm and a 16/32 mm fraction. 400 t of recycled aggregate was produced and used in the construction of a residential building and a parking garage. The fresh concrete was produced and delivered as planned, and the properties of the hardened concrete were satisfactory. However, high-water absorption of the recycled aggregate required careful control of the batching and mixing process. There were also practical problems of operational and logistic nature.

In 1989–1994 four full-scale test were carried out for the purpose of demonstration of crushed concrete and masonry as aggregates for the construction of new buildings (see Figure 4.18).

Figure 4.18 Four houses in Denmark built with recycled materials in 1991–1993. Industrial building (top left) is built with façade elements of masonry aggregate concrete (see Section 4.3.4). The three other houses are dwelling houses. (From Lauritzen, E. K. and Hansen, T., Recycling of construction and demolition waste 1986–1995. Danish Environmental Protection Agency, Environmental Review no. 6 1997.)

4.9.1 The Recycled House in Odense

As the most important part of the demonstration effort, three apartment houses were built in 1991–1994, predominantly with recycled materials. Figure 4.18 shows two of the houses. The purpose of the three projects was:

- To build functional residential buildings within the limits of a given budget, to the largest possible extent using recycled materials from demolition work
- To conduct investigations and practical experiments with recycled materials during construction

The demonstration projects were carried out as three independent projects with different owners, consultants and contractors. *The Recycled House in Odense* was the most complete and spectacular full-scale example of building with recycled materials.

It was the aim of the Recycled House project in Odense to study how much of the old materials from a condemned house nearby, built in 1890, could be recycled in a new 2.5 storey, 2 wing building with 14 cooperative non-profit flats in the centre of the town. A significant part of the new structure was built from recycled materials such as crushed concrete and crushed tiles used as aggregates in concrete, used timber for structures and joinery, and used bricks and roofing slates for façades and roof. The Recycled House in Odense exhibits in full-scale the technical state-of-the-art of recycling building materials.

It should be mentioned that all construction work on subsidised buildings must be awarded to the lowest bidder in a tender. This means that all specifications for, and control of, the work had to be described in tender documents and thus considered and defined before any practical experience was gained. To balance reliable specifications against the limited economy in the project was a great challenge for the architect and the engineer. Testing of the recycled materials were performed in parallel with the new building activities.

Cleansing whole bricks was one of the major problems. While working with the over 130,000 bricks on the project, it was revealed that the sorting of them was inadequate. Bricks from the front and back walls, outbuildings, stairways and partitions were mixed together. Nor had the bricks been graded according to quality or size after they had been cleaned.

It was planned to clean the bricks on the demolition site. However, it quickly became apparent that the extent of the cleaning process had been underestimated. The idea of processing the bricks on-site was to save energy in transporting them, thus in keeping with the spirit of the project. This point had to be abandoned, as it was necessary to establish several sites where the bricks could be cleaned. Since 2004, new inventions on cleaning bricks have been developed. Additionally, the whole bricks did not work out properly in the front walls of the recycled house. Due to improper sorting of the old bricks, many bricks in the front walls of the recycled house were damaged by frost scaling, and many more will be in the future.

Old wood turned out to be an easier material to reuse and recycle than old bricks. It became apparent that the old construction timber was of very good quality. It had a much longer growth period than timber harvested today, making it much stronger (see Figure 4.17). Construction of the rafters were carried out using processed recycled wood. Altogether, 2,000 running meters of rafters were used. The floors of the house are also processed old flooring. By turning the floorboards over and using the underside, after planning and introducing new grooves and tongues, they presented themselves to be exceptionally good. Due to the wood's seasoned age, they does not give. That made it possible to lay the floor quickly and precisely. It is important to store the wood properly so that it is not exposed to moisture. Flooring and timber were removed for further treatment. Nails and other metal objects were removed as far as it was possible before the wood was cut at a sawmill. Saw blades had to be sharpened more often, but this is something that can be remedied by the use of metal detectors.

The wooden windows were also recycled. The frames were planed to new measurements and a prototype was produced before construction in quantity was started up. Altogether, 68 double and treble glazed windows were produced for the project. The cupboard boards were manufactured from processed wood by a professional producer of new kitchens, and good results were achieved using recycled wood. The 14 main doors were made of recycled wood, while the 44 interior doors were old panelled doors. Together with the wooden floors and the masonry it creates a very rustic impression.

The slate roof was dismantled by selective dismantling. The laying of the 670 m^2 new roof went well.

Finally, concrete from a nearby air raid bunker was used with the crushed brick aggregate from the condemned building. However, most of the aggregate did come from a bridge, which had become superfluous because of the new Great Belt Link.

These and other similar experiments have clearly demonstrated that old crushed concrete used according to the instructions contained in the *Recommendations for the Use of Recycled Aggregate for Concrete in Passive Environment* is well suited as aggregate for production of new concrete in building construction. The specification was published by The Danish Concrete Association in 1989 (Danish Concrete Association 1989) and was the precursor for the new RILEM specifications for the same purpose. Usually, such RILEM specifications are precursors for new European CEN specifications, including *EN 206-2013 Concrete-Specification, Performance, Production and Conformity,* and *EN 12629 Aggregates for Concrete.* No problems were experienced while working with the approximately 275 m^3 of concrete in which the aggregate size varied from 4 to 32 mm. The new concrete was used for storey decks of the new building. All in all, 75% of the wood from the old building was used, as well as 80% of the slate and 50% of the old bricks, after contaminated and damaged pieces were sorted out.

The idea of recycling building materials stems from the problems with solid waste disposal. However, the recycled house project also shows that a building can achieve new architectural values through recycling because of the 'soul' in the old materials, which creates a pleasing impression, and it help to make people aware of problems and values within the context of recycling. People can talk with the residents and find out that it is possible to live in a better house than is usually built from new materials.

The demonstration projects, building houses with 80% recycled materials, are excellent examples of a cleaner technology and circular economy in which carefully selected dismantling has contributed to a reduction of the consumption of dwindling resources, and has had a minimum impact on the environment.

4.9.1.1 Experiences gained from the recycled houses

The recycled houses have made several valuable contributions. First, it was proved and fully documented that it is possible to build new houses for common purposes by the use of recycled materials following traditional building codes and legislation.

One of the most important experiences achieved from the project was the introduction of a set of norms for recycled materials. Building owners, engineers, architects and craftsmen alike must know their materials thoroughly when they plan and bid on such projects.

Another experience led to the conclusion that it is necessary to establish depots for storing recycled materials in order to achieve the flexibility and economy necessary for employing these materials. The right materials must be available in the right quantities at the right place and at the right time. When these conditions are met then a sound economy can also be attained in similar future projects.

4.9.2 Humbolt University, Berlin

CEMEX Germany, a private global manufacturer of building materials, has conducted a pilot project in Berlin, where recycled concrete was used throughout the construction process, together with the Berlin Senate (Senatsverwaltung für Stadtentwicklung und Umwelt). The purpose of this project was to show the way for the use of recycling concrete in public construction work.

Ca. 3,800 m^3 of concrete was delivered by CEMEX to the northern campus of the Humboldt University in Berlin. The builders chose to use a special building material, and the university's new laboratory and research building for life sciences was built almost exclusively of a concrete product produced by recycling units. The building was architecturally demanding, and CEMEX Germany developed for this purpose a special quality of C30/37 strength range, with a professionally manufactured 8/16 mm grain waste recycling unit. In order to create a consistent quality of concrete for the entire construction, CEMEX collaborated closely with the local company Recycling GmBh, which processes and certifies recycling units. CEMEX and the developer encountered several expected and unexpected challenges (such as material availability and the technical properties of recycling concrete). All challenges were resolved and, as expected, the project has contributed valuable experience in the production and use of resource-efficient building materials (Danish Competence Committee for Waste and Resources [DAKOFA] 2015).

4.9.3 RAC projects in Germany

With the aim of promoting the use of reusable concrete in buildings, the German environmental authorities (Bundesministerium für Umwelt, Naturschutz und ReaktorsicherheitIfeu), together with several public and private partners, financed a demonstration project *Pilot Project RC Concrete*. The project included 14 individual projects in Rhineland Palatinate and Baden-Württemberg, where recycled concrete was used for construction projects as Type 1 and Type 2 mixtures:

- Type 1: 90% recycled aggregates
- Type 2: 30% masonry, 70% recycled aggregates

As part of the project, the first house of recycled concrete was completed in Ludwigshafen-Friesenheim in March 2010. The positive results of the pilot project have increased the use of recycling concrete, especially in the Stuttgart area. Here, several concrete factories have added recycling concrete to their production since 2011, and the material is widely used for buildings.

4.10 GLOBAL VIEW OF THE RECYCLING MARKET

As CDW materials make up between 25% and 50% of all waste generated by human activities, and are dominated by concrete materials, it is relevant to take a global view of the needs and opportunities for recycling of CDW materials substituting mineral resources (Lauritzen and Hansen 1997).

Three main factors must be considered when evaluating the prospects for recycling of mineral CDW waste at any given location, as illustrated in Table 4.12. In order of importance the three factors are:

- Population density
- Natural raw material deposits
- Level of industrialization

Table 4.12 Table summarizing the main factors affecting the prospects for successful recycling of CDW (Lauritzen and Hansen 1997)

Area code	Population density	Natural raw materials deposits	Level of industrialization	Example	Prospects of successful recycling of CDW
I	High	Adequate	High	Many cities in EU, United States and the Far East, e.g. Hong Kong (China), Copenhagen (Denmark)	+++
II	High	Adequate	Low	Many megacities in the Third World: Mexico City (Mexico), Jakarta (Indonesia)	+++
III	High	Scarce	High	Amsterdam (The Netherlands)	++++
IV	High	Scarce	Low	Dacca (Bangladesh), Calcutta (India), Shanghai (China)	++++
V	Low	Adequate	High	Rural Scandinavia	++
VI	Low	Adequate	Low	Rainforests and mountain regions in developing countries	+
VII	Low	Scarce	High	Kuwait Abadan/ Khorramshahr (Iran)	+++
VIII	Low	Scarce	Low	Steppelands or sandy deserts or tundras in developing countries	+

Rating from '+' (doubtful) to '++++' (profitable).

Other important, but unpredictable factors are wars and natural disasters such as earthquakes, hurricanes and floods, as described in Chapter 6.

Population density is by far the most important factor determining the viability of recycling CDW materials. There is no economic basis for large-scale recycling in sparsely populated parts of the world. High population density is required in order to ensure an ample and steady supply of wastes, which can serve as raw materials for recycling industries. High population density is also associated with new building activities, which provide markets for recycled products. In addition, there is a scarcity of landfill sites in densely populated areas, and for environmental reasons, there is usually strong opposition to incineration plants.

Scarcity or difficult access to natural raw materials favour recycling of CDW at a high technological level. Abundance and easy access to natural raw materials do not preclude recycling of CDW per se, but for economic reasons it usually dictates activities at a lower technological level. To some extent, the level of industrialization affects the need and desire of a society to recycle CDW materials. In densely populated developing societies, sanitary and social reasons dictate that the amounts of waste, which are brought to landfills, should be reduced as much as possible. In industrialized parts of the world, there are additional and, to some extent, more subtle aesthetic, environmental and economic reasons why landfilling and incineration of waste should be reduced to a minimum. One obvious way of alleviating some of the problems is to sort out and recycle the heavy and bulky CDW from the general waste stream. It is evident that in principle it should be attempted to make methods employed in the collecting, sorting and processing of CDW in developing countries as labour intensive as possible, and as mechanized as possible in industrialized parts of the world. However, in practice it turns out that the most labour intensive methods are not always are the most economical, even in the least developed societies. On the other hand, suitable mechanized methods of handling many kinds of CDW are not yet available. Recycling plants must often resort to laborious handpicking even in industrialized societies. This is expensive and associated with health and safety hazards.

Practical methods and systems of collecting, sorting and processing of CDW need further study and practical development both from economic and technical points of view, as mentioned in later in Chapter 5 and illustrated by Table 4.12, which shows a macroeconomic model of integrated resource management and total costs.

The area codes are explained as follows.

Area code I in Table 4.12 is exemplified by Hong Kong, which has a high population density and a high level of industrialization in many commercial sectors. However, because of easy access to raw materials from mainland China, high-grade processing of CDW waste into new building materials is not economically viable. The main problem in Hong Kong is to achieve a sanitary and economically satisfactory disposal of waste. To a large extent, this is accomplished by the sorting of rubble from the general waste stream. After crushing, screening and washing at central processing plants, most rubble is used as fill for land reclamation purposes.

Area code I is also exemplified by Copenhagen in Denmark. Copenhagen has one of the most advanced collecting and treatment schemes for CDW in the world. Yet, the fact that there is ample supply and easy access to natural raw materials makes it economically unviable to process CDW into higher-grade building materials. Thus, almost all CDW go into road construction and maintenance where it substitutes up to 10% for natural sand and gravel.

Area code II covers megacities in the Third World, which are close to raw materials sources but where access is difficult, typically due to permanently congested traffic conditions. Jakarta in Indonesia and Mexico City are examples in point.

Area code III is exemplified by Amsterdam in The Netherlands. There is extensive generation of CDW and intensive new building activity. The population density is high. There are almost no accessible natural raw materials and the environmental consciousness among the population is high. As a consequence, The Netherlands is one of the leading nations in the world for what concerns high-grade recycling of waste. Crushed CDW rubble is used as concrete aggregate as a matter of routine, and advanced equipment is being developed in The Netherlands for the processing of various kinds of waste into higher-grade products.

Area code IV is exemplified by Dacca in Bangladesh with a population of 15 million and a very low level of industrialization. The city is located far from natural raw materials deposits; access is very difficult and new materials are expensive. This is illustrated by the fact that clay is burned to bricks within the city limits using natural gas as fuel. The bricks are then crushed and used both as aggregate in new concrete and for road construction purposes. Located in a river delta, the region south of Dacca is notorious for recurrent disastrous floods, which require frequent reconstruction of roads. Under such conditions, crushed and screened CDW rubble would easily find a market. At the same time, elimination of CDW waste from the general waste stream would help to alleviate some of Dacca's urgent landfill problems. Similar conditions exist in many other large cities in the developing world, particularly cities located in or near river deltas. Examples are Calcutta (Ganges) in India and Shanghai (Yangtze) in China.

In parts of the world identified by area codes V, VI and VIII the concept of recycling is primarily associated with interests in environmental protection. Traditionally, demolition waste is tipped in nature due to convenience and availability of space. Only larger demolition projects, for instance in connection with reconstruction after disasters and wars, will provide reasonably economic options for the recycling of CDW.

Area code VII is exemplified by Kuwait, where most natural raw materials are imported from the United Arab Emirates thousands of kilometres away. This makes the cost of materials extremely high. Economic and technical feasibility studies of recycling of CDW in Kuwait, prepared by the Danish consulting engineering companies Rambøll and DEMEX in 1990–1991, have clearly demonstrated that recycling of CDW in Kuwait would be profitable.

Chapter 5

Integrated CDW management

> The integrated management of buildings and building materials in urban area, called the City Concept, is a holistic way of thinking and a specific way of doing.
>
> *The IRMA Project, 2006*

5.1 INTRODUCTION

The building owner who intends to remove existing buildings and construction considers the materials from demolition work as waste to be handled for the lowest possible price. The building owner who needs resources for the construction of new structures might see an advantage in using the waste materials as resources in the new structures. The overall outstanding question is how to convert unwanted CDW to demanded resources substituting natural resources and building materials. Referring to the discussion of the waste/resource dilemma in Chapter 1 and the transformation of buildings and materials in Chapter 2, this chapter deals with integrated resource management with the fulfilment of mutual needs with shared values and synergy. The key elements of resource management are demolition, described in Chapter 3, and recycling described in Chapter 4.

The preconditions for integrated CDW management must be viewed through two perspectives: the waste perspective and resource perspective. Traditionally, CDW management is a part of the waste management sector. Due to the amount of waste and content of pollutants, the EU and many countries consider CDW as a prioritized waste type in line with food waste, plastic waste and electronic waste. Therefore, it is mainly the waste and environment sectors which take responsibility of CDW management and the development of steps to 3R (Reuse, Recycling, and Recovery of CDW). The CDW management is more or less neglected by the building and construction sector, except the dedicated demolition contractors. The demolition contractors form a specialized minor sector of the building and construction sector. The waste perspective dominates most international literature on demolition and recycling. The research and development of 3R technologies and instruments are based on the need of getting rid of waste.

Historically, we have always reused bricks and stones. Today, we are familiar with recycling asphalt, bricks, gypsum and crushed concrete. Reuse of windows, doors and building materials take place and recycled items can be purchased on the Internet market. Research and development together with demonstration projects in Europe during the past 30–40 years have provided many opportunities for the economic and environmental recycling of CDW (see Chapter 4). However, we are only interested in cheap low-quality recycling and easy solutions. High-quality recycling of concrete, as aggregates in new concrete, is not generally implemented in the building sector despite the fact that the technical preconditions exist.

On the other hand and from a resource perspective point of view, the resources transformed from CDW are needed. Referring to the UN 17 Sustainable Development Goals 2030 and the EU initiatives on circular economy, climate change and resource efficiency, the building and construction sector has taken notice of the issue of saving resources. The points given for DGNB, BREAM and LEED or other certifications related to planned deconstruction, recycling and reduction of resources are rather low, and the contractors in most of the industrialized world still obtain cheap sand and stones from nature.

The overall advantage in recycling and utilization of CDW is improved sustainability and circular economy, characterized by economic profit and saving resources for the building owner who needs to get rid of the CDW, as well as the building owner who need resources. Expressed in popular terms, to create a 'win-win' situation, one needs a mutual fulfilment of 'matching' the respective needs.

In this book, integrated CDW management means management of CDW materials with respect to the optimal, sustainable and resource efficient utilization of the recycled materials substituting natural resources. The integrated CDW management is project-oriented and focused on greater building and construction projects, or urban renewal/rehabilitation programs including transformation of existing (old) structures into new structures. Figure 5.1 and Table 5.1 illustrate the basic business model of integrated CDW management compared with traditional CDW management. The CDW management should be considered as an integrated element of the overall logistic management of urban development projects including all kinds of products and resources.

5.2 INTEGRATED CDW MANAGEMENT SCENARIOS

Based on the life cycle transformation presented in Chapter 2, three scenarios of transformation of old buildings into new buildings are presented in Figure 5.2.

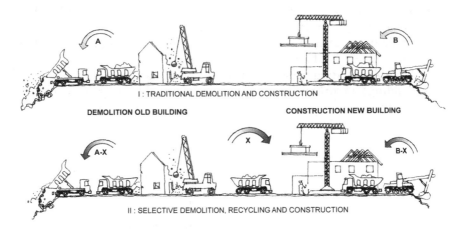

Figure 5.1 Principle of business model of integrated CDW resource management. The top figure illustrates a demolition project and a construction project without any synergy. The bottom figure shows exploitation of the recycled CDW materials from demolition as resources for the construction of a new building. (The model, developed by the author, is presented in numerous papers and publications, including Lauritzen, Erik K. and Hansen, Torben, Recycling of construction and demolition waste 1986–1995. Danish Environmental Protection Agency, Environmental Review no. 6, 1997 and Lauritzen E. K. Emergency construction waste management, Safety Science 30 (1998).)

Table 5.1 Principles of business model of integrated CDW management compared to traditional CDW management, illustrated in Figure 5.1

Type of CDW management	Cost/CO$_2$
I. Traditional CDW management, demolition and construction:	
Handling A tonnes CDW, transport, fees, disposal, etc.	I.A
Providence of B tonnes raw materials, production, transport, etc.	I.B
Total Cost/Load	I.A + I.B
II. Integrated CDW management, demolition, recycling, construction:	
Handling A − X tonnes CDW, transport, fees, disposal, etc.	II.A − X
Providence of B − X tonnes raw materials, production, transport, etc.	II.A − X
Transformation, crushing, qualification of X tonnes recycled materials	X
Total Cost/Load	(II.A − X) + (II.B − X) + X

Preferences:

I.A + I.B > (II.A − X) + (II.B − X) + X → Integrated CDW management to be preferred

I.A + I.B < (II.A − X) + (II.B − X) + X → Traditional CDW management to be preferred

- Single-building transformation from the life cycle of an old building to the life cycle of the reused building, Figure 5.2a
- Single-building transformation from the life cycle of an old building to the life cycle of a new building, Figure 5.2b
- Multiple-building transformation involving transformation of several buildings in urban development from an old city to a new city, Figure 5.2c

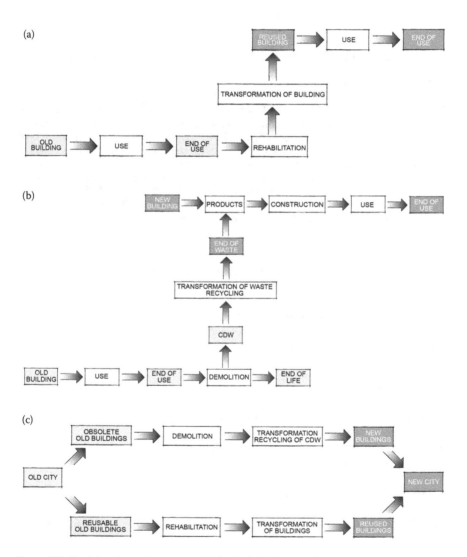

Figure 5.2 Models of transformations: (a) Single-building transformation, reuse of building; (b) Single-building, demolition of building; (c) Multi-building transformation of urban area.

In case of demolition, reconstruction and construction of a single or small number of structures and buildings, performed under one or a few numbers of contracts, we call it a *single-recycling scenario*. The related contracts are performed in a shorter period, typically 1–5 years. In case of demolition, reconstruction and construction of several structures under several independent contracts, we talk about a *multiple-recycling scenario*. Urban development programs, transformation of industrial sites and harbour areas into housing estate over a longer period, for instance 5–25 years, are considered multiple-recycling. The development of the Olympic Park for the London 2012 Olympic Games, and transformation of Rotterdam Harbour and Carlsberg Brewery, presented later in this chapter, are typical examples of multiple-recycling projects. Depending on the project and the space of the site, we distinguish between *on-site recycling* and *off-site recycling*, as characterized in Table 5.2.

Besides CDW waste reduction, recycling and efficient use of resources, the objective of integrated CDW management is to create and implement synergy and exploit the win–win situation between the demolition contracts and the construction contracts, as illustrated in Figure 5.1.

Therefore, the following preconditions for integrated CDW management must be met:

- Basic understanding of the need for substituting natural resources with recycled waste materials
- Dialogue based on mutual, equal understanding between the waste sector and the building and construction sector
- Matching demolition contracts and production of CDW with new building and construction contract needs for resources
- Establishment of a professional and viable market for recycled CDW resources

Table 5.2 Types of transformation, recycling scenario and types of recycling

Transformation	Recycling scenario	On-site recycling	Off-site recycling
Single-building transformation *Reuse of building* Single-building transformation *Demolition of building*	Single-recycling	Replacement of one or few buildings. Partial demolition, total demolition and construction on the same site	Separate demolition and construction contracts and separate sites. Outside supply of recycled materials
Multiple-building transformation *Reuse and demolition of buildings*	Multiple-recycling	Mixed demolition, renewal and construction projects in a major area	Separate demolition and construction contracts and separate sites inside the area. Supply of recycled materials outside the area

In principle, CDW management is very basic and simple. Planning and management of CDW handling calls for systematic thinking and simple computer programs. It is important to be familiar with the processes, the links and logistics related to demolition, as well as recycling and construction. Furthermore, the overview of the challenges, opportunities and risks is of crucial importance.

5.3 SINGLE-RECYCLING PROJECTS

A single-recycling project deals with a single-demolition project and a single construction project. Depending on the actual circumstances, the demolition work and the construction work might take place on the same site under one or two separate contracts, or on separate sites under separate contracts.

The traditional approach of a single-recycling project is characterized by the following:

- The demolition contract and the construction contract are planned and executed separately. The demolition contractor demolishes the old building and structures and delivers the site cleaned of all rubble and pollution to the owner. Normally, the design of the new building is based on an empty and cleaned site, which is handed over by the owner. While all efforts on planning and design are focused on the construction project, very little attention is paid to the demolition contract, which is quite understandable considering the budget of the projects
- Because of the need for the old building and shortening the time from the end use of the old building to completion of demolition and construction work of the new building, the time schedule of demolition work is often critical and tight
- Sometimes, the demolition contract is merged into the construction contract, which opens up opportunities, flexible time schedules and overlapping of demolition and construction activities. However, the building owner might often prefer a separate demolition contract, because of the specialized demolition work
- The completion of demolition work is expected before the start of the construction. This means that the handling of CDW also must be completed before the start of the construction
- Usually, the planners and designers of the new building do not pay any attention to the opportunities for the use of recycled materials from the demolition work, potentially providing building materials and resources needed for the new construction. On the other hand, the demolition contractor does not know the opportunities for selling recycled materials to the construction project

The optimal approach with respect to CDW and resource management includes following:

- The potential of recycling and reuse of demolition waste materials should be assessed at an early stage of the planning and design of demolition
- The opportunity of substituting natural resources and building materials in the new building should be assessed at an early stage of the planning and design of the new construction and before tendering of the demolition project
- The assessed potential of recycling of reuse of demolition materials should be matched with the assessed potential of opportunities for using recycled materials in the new building
- The options for recycling should be incorporated in the tender documents for both contracts, and the contractors should have the necessary time to assess the advantages and benefits in their technical and financial proposal for the contracts

5.3.1 Planning and management of single-recycling projects

5.3.1.1 Initial considerations

The single recycling project starts with the building owners need for a new building or structure on the site, and he/she initiates the building program based on his specific requirements. Depending on his/her and the investors ambition with respect to sustainability, circular economy and CO_2 reduction, he/she will consider the opportunities for certification according to LEED, DGNB, BREEAM or other certification schemes. In the perspective of a possible certification, the building owner will meet requirements for the materials for the building project and requirements for the transformation (demolition and recycling) of the materials by the end of the building's life cycle.

Depending on the quality of the existing old building and the quality of materials, part of the buildings or specific items of historic/cultural value might be designed for reuse in the new building, for instance granite stones, windows, doors, timber and bricks. In case of the presence of suitable amounts of concrete, asphalt and bricks applicable for utilization in the new building project, detailed assessment are needed.

5.3.1.2 Assessment of the potential of recycling and utilization

Referring to the EU package on circular economy (EU Commission 2015), it is the intention of the EU to introduce requirements for mapping the recyclable/reusable inventory of a building before demolition. In this case, the priority of the assessment is to find materials applicable for the new building project, and also to assess materials applicable for other purposes. Mapping of recyclable materials should be based on detailed inspections of the building and assessments of type and amount of materials. The basic principle of the assessment is starting from the roof and ending with the foundation and surroundings. Referring to Chapter 4, a wide range of possibilities of recycling and reuse of building materials are described. The technical feasibility of the reuse of stones, bricks, timber, doors and windows and the recycling of asphalt, crushed bricks and concrete are assessed. However, when it comes to the economic and environmental feasibility with respect to the common construction of buildings, only limited possibilities of reuse/recycling are realistic and competitive with the building materials and resources on the market. Therefore, the assessment of the potential for reuse/recycling of a house before demolition must be based on a very critical understanding and practical perception of realistic needs of resources. Referring to Chapter 4, the following should be noted during the assessment of recycling opportunities:

- Some roofing materials can be reused. However, roofing materials, such as tiles, bricks and slates are normally not applicable for reuse because of wear. The dismounting of roofing materials must be done by hand. Many different types of roofing materials are found, which makes it difficult to find costumers
- Windows and doors are seldom applicable in new buildings because of wear and technical requirements. Reuse of windows and doors is common in the private market
- Usually timber and wooden flooring are reusable. Often, you can find a high demand for timber and flooring, because of the high quality of old wood. However, the removal, cleansing, extraction of nails and fittings, etc., is time-consuming handwork. In some countries CDW wood materials are primarily shredded and used for various purposes or incinerated
- Except special items of culture or decorative interest, metal and iron in a building, including heating installation, machines and construction steel are normally scrapped and recovered. In case of reasonable amounts of current types of construction steel profiles, reuse might be relevant. During demolition work in poor and developing communities, reinforcement bars from concrete elements are extracted and recycled
- Bricks are reusable if they can be separated and cleansed without damages. It must be noted that only bricks from buildings made with hydraulic lime mortar, or mortar with a very low content of cement, can be separated. Masonry made by cement mortar-bounded bricks, typical of constructions after the 1950s, is very difficult to disintegrate. There is a high demand for reused bricks to repair old masonry buildings

- Recycling of concrete for coarse aggregates in road constructions, or basement of surfaces, is common. Recycling of concrete for aggregates in new concrete, RAC, can be used for recycling of concrete pavement and roads. RAC is applicable for passive environmental class concrete
- Gypsum, glass and mineral wool can be reused, recycled or recovered
- Pavement tiles and stones can be reused. However, only granite stones are preferred for reuse
- Major asphalt surfaces can be recycled
- Reuse of sand and soil might be relevant for fill and landscaping
- Installations and inventory such as heating, water and sanitation (HWS) objects might be reusable. However, this is normally not relevant in professional construction of new buildings because of technical requirements, change of trends and risk of poor functionality or pollution. Reuse of electric installation and HWS objects are often seen on the private market

5.3.1.3 Assessment of opportunities for substituting natural resources and building products with recycled materials in the new project

Based on the requirements of the new building, the assessment of opportunities for substituting natural resources and building materials from the existing building comprises following:

- Special objects of interest. In case of replacement of an existing building with special historic/cultural/architecture qualities, the design of the new building might be inspired by the desire to reuse parts or materials of the old. This might lead to the requirements in the demolition contract to remove the objects before or during the demolition
- Reusing granite stones, tiles, doors, windows, timber, etc., in the new building or in landscaping and decoration outside the building
- RAC replacing concrete of passive environmental class in the structure, e.g. foundations, slaps and parking structures. Initially, the assessment focuses on the possible match with respect to amounts of concrete in the old structure and then the need of concrete in the new structure, including logistic opportunities of recycling concrete on the site
- Crushed aggregates as sub-base for new roads, footpaths and surfaces around the new building
- Crushed concrete and masonry for fill and landscaping around the new building and fill around pipes, heating transmissions, etc., in the ground
- Crushed concrete and masonry for roads, surfaces and interim structures on the building workplace during the construction

5.3.1.4 Matching – Logistics and timing

The logistic planning and management is one of the most important and decisive issues of the recycling project. Usually, the demolition work and CDW production start some time before the need for materials in construction of the new building. Opportunities for temporary storage of materials, recycling and stockpiling of crushed materials depends on the space available at the site. The optimal site requires space for handling, sorting, crushing and stockpiling the various fractions of materials during the recycling processes. In case of lack of space, opportunities for available sites must be searched for in the neighbourhood. The distance to the available recycling site influences the economy, environment load and the overall feasibility of the recycling project.

The demolition contractor must see an economic advantage in his handling of the demolition waste, with respect to specific recycling opportunities, as a positive alternative to traditional CDW handling. On the other side, the building contractor must see an advantage to provide recycled materials prior to resources from the traditional market. However, *matching of timing and logistics is a precondition for the optimal recycling project!*

5.3.1.5 Indicators

For each recycling scenario, the following key indicators should be assessed and discussed:

- Increase in the number of reusable buildings and amount of recyclable waste (more than 90% of the total building waste)
- Substitution of natural resources by recycled materials (percentage of the consumption of natural quarry materials)
- Reduced transport and environmental impact (measured in tonnes/km and amount of CO_2 emission)
- Improved indoor climate and reduced risk to occupants and workers (measured in hazard probability and hours/days of absence)
- Total cost of the different demolition and CDW management scenarios, based on economic models and field monitoring of actual projects
- Total cost of different building rehabilitation scenarios, based on monitoring and cost assessment

5.3.1.6 Ownership responsibility

Usually, in demolition contracts the contractor will get the ownership of all building materials, installations, left inventory and goods in the buildings comprised in the contract, except for items marked as the property of the building owner. In case of a recycling project, where CDW are recycled and designated for a new building, the transfer of the materials from the demolition to the construction must be legalized and commercialized. We distinguish between two different models:

Appointment between the building owners

The building owner of the demolition contract makes a deal with the building owner of the construction contract on providing the recycled materials to the construction project in accordance to specific terms.

This presupposes:
- That the demolition contractor must treat and stockpile the materials on behalf of the building owner. For instance, it could be downsizing, sorting, crushing, sieving, screening and stockpiling in specific types and fractions
- That the construction contractor must accept recycled materials provided by the building owner

This model has the advantage that the recycling processes and transfer of materials are secured and controlled by the building owners. On the other hand, the contractors do not have the freedom of choice, which might make the total recycling budget higher.

Appointment between the contractors

The demolition contractor sells recycled materials to the construction contractor.

This presupposes:
- That the building owner accepts the option of the demolition contractor to sell recycled materials to the building contractor
- That the building owner of the construction contract accepts the construction contractor's option of using recycled materials

This model gives the two contractors freedom to negotiate and make their appointment. However, the model contains the risk that the two contractors cannot agree on the recycling business.

Generally, the model mentioned last is recommended in order to keep the building owners out of the responsibility of the transaction between the contractors. However, the greatest chance of success presuppose a preceding dialog and a mutual understanding of the objective of the recycling process. In the case where the demolition contract and the construction contract are under the same building owner, or the demolition contract is a subcontract under the construction contract, there should be no problems.

5.3.1.7 Responsibilities of quality and purity of the recycled materials

According to EU regulation, reuse and recycling of CDW requires that materials are clean without hazardous materials. The threshold values of contamination is up to national standards of the EU member states, and some states approve recycling of minor polluted materials for reuse and recycling, under controlled conditions, without the risk of polluting the environment, especially the groundwater. The demolition contractor is responsible for cleansing the demolition site before demolition, and he is responsible that only clean materials are recycled or reused. In response to the building owner's and contractor's requirements to ensure the purity and the avoidance of the risk of pollution of the new structure, it is recommended to set up a programme for quality assurance and testing the purity of materials.

The EU directive on building materials and the CEN and EN standard rules establishes standardization and qualification of materials for new materials. The *Construction Product Regulation* (EU Commission 2011) describes the requirements for CE marking of building materials. Similar rules for recycled materials do not exist even if they comply to the EOW criteria (Saveyn et al. 2014) according to the EU Waste Framework Directive (EU commission 2008). In the case where the reused materials fully comply with the technical specifications for new materials there is no problem. In some cases, the materials comply with new materials with respect to functionality, for instance timber and bricks, but have another appearance or other discrepancies (such as holes in timber from nails and bolts other colours and shapes of bricks) the building owner might nevertheless accept the materials. In Europe, there is ongoing work on standardization of reused building materials.

Reuse of doors and windows do often meet requirements for fire, climate, energy and noise transmission, which prevent reuse in new buildings. The authorities must approve the technical quality. Building owners and architects might prefer reused building materials for aesthetic and cultural reasons. However, the technical specifications must be fulfilled, or they might take the risk of possible disadvantages. Reused doors and windows of good quality might be upgraded with extra layers of wood or glasses in order to comply with standards for new materials. Generally, in Europe, the reuse of building materials is mostly applicable to private homes and cottages, and the market of reused materials addresses private customers.

5.3.2 Decision of single-recycling projects

The decision to carry out the demolition project and the construction project as one contract on the basis of integrated CDW management or as two separate contracts on the traditional basis without recycling materials, depends on a systematic analysis of the actual opportunities and the above-mentioned considerations. A simple life cycle assessment (LCA) comprising the most important parameters will form the background, for instance: economy, risks, number of worksites, energy and CO_2 emission.

Referring to the introduction of this chapter and Figure 5.1, we are discussing two scenarios:

- *Traditional resource management* with supply of natural resources and external disposal of CDW
- *Integrated resource management* with the greatest possible local recycling of CDW substituting natural resources

The integrated resource management scenario is simple, but it is not practiced to a great extent for many different reasons (Figure 5.3). Recycling of bricks and concrete is usually done by removing the materials from the demolition site and handing them to receiving facilities, or recycling takes place entirely in the vicinity of the demolition site, that is demolition contractors usually have to pay to get rid of recyclable bricks and concrete waste. They have no financial gain in recycling. Today, construction projects and demolition projects are offered on the assumption that the handling and disposal of the CDW is a cost. Apart from large demolition projects with large concrete volumes, it has not been common to offer demolition projects to exploit the resource potential from the large fractions of concrete and bricks.

An example of the methodology for the assessment of whether to use traditional resource management or integrated resource management is presented in Table 5.3.

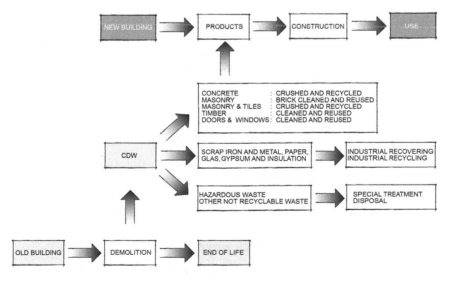

Figure 5.3 Process diagram of a typical single-recycling project from an old building to new building.

Table 5.3 Methodology of calculation cost and CO_2 emission

Traditional CDW management	Cost $,€..	CO_2 kg	Integrated CDW management	Cost $,€..	CO_2 kg
A. CDW handling:			**A. CDW handling:**		
• Demolition	...	...	• Selective demolition	...	...
• Transport of CDW	...	...	• Transport of material	...	...
• Handling and	...	...	for recycling	...	...
disposal of CDW			• Transport of remaining	...	...
			CDW		
			• Handling of remaining		
			CDW		
B. Supply of natural			**B. Supply of natural**		
resources:	...	...	**resource:**	...	...
• Extraction from	...	...	• Extraction from quarry	...	...
quarry or sea	...	...	or sea	...	...
• Crushing and sorting			• Crushing and sorting		
• Transport to			• Transport		
construction site					
			X. Recycling material:		
			• Transport cost	...	...
			• Treatment/crushing,	...	...
			sorting, sieving	...	...
			• Testing	...	...
			• Packaging and delivery		
Total cost and CO_2 emission	$\sum$...	$\sum$...	**Total cost and CO_2 emission**	$\sum$...	$\sum$...

The explanation for the lack of recycling on-site is primarily that it is often problematic to match the availability of waste materials with the need for raw materials in terms of time and place. Recycling on-site requires purpose, planning and preparation time. The explanation is related to traditions, cultures, and often a very limited overall view. Architects and engineers think more in terms of new constructions and new materials then about the savings and recovery of resources from demolitions of old worn structures.

5.3.3 Implementation of single-recycling projects

Usually, the demolition work and construction work are implemented as two individual contracts in accordance to the normal work standards and regulations, comprising elements listed in Table 5.4. However, the demolition contract might be subcontracted under the construction contract.

5.3.4 Risk and sustainability perception

As shown in Figure 5.4, the building owner and the contractors in a single recycling project will experience a jointly agreed economy, but they will have a different perception of profit and risk. In some projects,

Table 5.4 Elements of demolition and construction projects

Elements of the demolition contract	Elements of the construction contract
• Planning and design of demolition • Pre-demolition assessment • Tendering, requirements on recycling • Contract • Establishment of worksite • Cleansing • Selective demolition • Waste handling and recycling, providing resources to new construction site • Finishing • Quality control and documentation, including CDW handling reports	• Planning and design of construction work • Assessment of resources • Tendering, requirements of substituting natural resources with recycled resources • Contract • Establishment of worksite • Earth and foundation work • Construction work • Supply of resources and recycled materials • Landscaping • Finishing • Quality control and documentation, including building reports

the building owner and the contractor agree on specific sustainable goals based on agreed preconditions. However, usually important elements of sustainability are neglected, such as effects of transportation (CO_2 emission, energy consumption, dust, noise, vibration, threat to animals and wear of roads, etc.), consumption of resources, use of land and other kinds of externalities. There is much talk about sustainability and

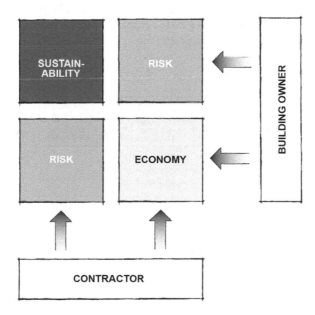

Figure 5.4 Differences of views on risk. The economy is agreed. The contractor and the building owner have different views of their own business risk. Their views and perception on susatainability including recycling and CO_2 emission are uncertain.

Corporate Social Responsibility (CSR) in larger and high profile projects. However, when you get into the construction site and look at the project context, the implementation will focus on production, time limits, safety and economy – expectations of success and the fear that it will go wrong. This applies not least to the many different risks associated with any project. In projects with elements of recycling of CDW, there is always a risk that the expectations and ambitions of the products are not met. Compared with the risk of defects in the delivery of natural materials, the risk of defects in recycled materials must not differ materially. No builders run any unnecessary risk by purchasing recycled materials unless they are sufficiently compensated economically. Please refer to risk management in Chapter 3, Section 3.7 'Demolition Methodology'.

5.3.5 Case story, reuse of bricks

Reuse of bricks is a high priority initiative of the waste management strategy of the Danish government, supporting sustainable development and circular economy. In connection with the extension of Katrinedals School in Copenhagen, the City of Copenhagen has decided to build the new buildings with façades of reused bricks from the demolition of a building, Building no. 13, Bispebjerg Hospital. Figure 4.15 in Chapter 4 shows photos of the processes of demolition, transformation and building with reused bricks. Figure 5.5 presents the processes graphically. Table 5.5 discloses the main figures of the business case and CO_2 emissions. The project took place from September 2015 to October 2016. Approximately 160,000 bricks were reused (Copenhagen Municipality 2017).

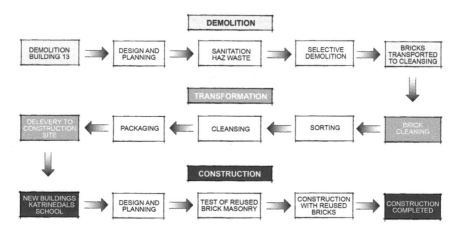

Figure 5.5 Case story of single-recycling project. Processes for reuse of bricks from the old hospital building to the new school buildings, refer to the photos in Figure 4.15.

Table 5.5 Business case referring to Figure 4.15 and Figure 5.5

Traditional CDW management	Cost €	CO₂ kg	Integrated CDW management	Cost €	CO₂ kg
A. CDW handling			**A. CDW handling:**		
Transport			**Transport**		
5,600 t masonry to landfill site, 40 km			5,600 t masonry to cleansing station, 27 km		
160 truckloads, each 35 t, 40 km, 3 hours per tour,			160 truckloads, each 35 t, 27 km,		
80€/hour:160 × 3 × 80€	38,400		2 hours per tour, 80€/hour: 160 × 2 × 80€	25,600	
Tipping fee			**CO₂ emission**		
6€/t: 5,600 t × 6€	33,600		160 truckloads, each 3.5 kg CO₂/km × 27 km		15,120
CO₂ emission			**Payment to demolition contractor**		
160 trucks, each 3.5 kg CO₂/km × 40 km		22,400	160.000 bricks, 7 Cent/brick	11,200	
B. Supply of natural resources:			**B. Supply of natural resource**		
160,000 bricks, 0,9 -€/brick, incl. transport	144,000		No extra bricks needed	0	0
CO₂ emission, transport 325 km.			**X. Recycling material:**		
11 trucks × 3.5 kg CO₂/km × 325 km		12,513	**Recycling**		
			Sorting, cleansing, packaging and delivery on construction site, 25 km		
			160,000 bricks each 0.85€/pc	136,000	
			CO₂ emission		
			11 truckloads, each 3.5 kg CO₂/km × 25 km		963
			Saving CO₂ of production of reused bricks compared to new bricks		
			−0.40 kg CO₂ per reused brick × 160.000		−64,000
Total	**216,600**	**34,913**	**Total**	**172,800**	**−47,918**

Source: Copenhagen, Municipality of Copenhagen, Genbrug af mursten. Erfaringer fra nedrivning af bygning 16 på Bispebjerg Hospital og genbrug af mursten til nybyggeri på Katrinedal Skole, Vanløse, 2017 [In Danish].

Note: Reuse of bricks from the demolition of the hospital building for construction of a new school building in Copenhagen, comparison of traditional and optimal (actual) resource management with respect to cost and CO₂ emission.

The single recycling project on reused bricks was based on the following conditions and presumptions:

- The demolition contract, including careful recovery of masonry and bricks destined for reuse, should not be more expensive than traditional demolition without any intention of recycling bricks
- The budget of the construction project should not be extended because of the reuse of bricks compared with new bricks
- The demolition project and the construction project should match with respect to delivery of qualified and approved tested bricks on required time and in sufficient amount

Table 5.5 presents comparison of the main figures of cost and CO_2 emissions between a traditional resource management scenario and the integrated (actual) management scenario based on following data:

- The figures of the estimated traditional CDW management was based on delivery to a nearby crushing and recycling plant, transport distance 40 km, and tipping fee 6€ per ton
- Supply of new 160,000 brick from the brickwork, 325 km to the construction site in Copenhagen. Price including transport estimated to 0.9€ per brick
- The demolition contractor *P. Olesen and Sons* transported 5,600 t, 160 truckloads each 35 t, of masonry to the recycling company. *Old Bricks*. Transport distance 27 km, return trip 2 hours. Transport cost 80€ per hour. The demolition contractor achieved a fee of 7 cent per reused brick (0.5 DKK)
- The recycling company produced 160,000 reusable bricks, 384 t, 11 truckloads, which were packaged and delivered on construction site; transport distance 25 km. Cost incl. packaging and transport 0.85€ per brick. The remaining part of the 5,600 t delivered from the demolition contractor was handled and reused for other purposes without significant economic effect on the case story
- For CO_2 emission calculation
  - Saving of 0.4 kg CO_2-eqv. per reused brick according to (Møller et al 2013)
  - CO_2 emission 3.5 kg per km transport of 35 t truckloads (Copenhagen 2017)

Key experiences:

- The business case showed a traditional management cost of 216,000€ compared to an integrated management cost of 172,800; 80% of the traditional cost. There is only little difference between the price of new bricks and the price of reused bricks. While the prices of handling the masonry from the demolition vary significantly

- The significant savings of 47,918 kg CO_2 compared to the emission of 34,913 kg CO_2 demonstrates clearly the advantages of recycling. The reduction of CO_2 emission from the production of recycled bricks is extremely lower than the production of new bricks. Also, the CO_2 emission from transportation during the integrated management scenario is significantly lower than transportation during traditional management
- Generally, old bricks of normal quality are technically suitable for reuse presupposing that the masonry is made of mortar without (or a low content of) cement, and the bricks can separate without damages
- The demolition contractor was paid 0.5 DKK (7 Cent) per reused brick, which represents an important income, compared with the expenses of traditional demolition and CDW management. This income should also be to the advantage of the building owner, reducing the cost of the demolition work
- The recycling company delivered the 160,000 bricks to the building contractor according to the agreed technical specification and test procedures, and with a unit price at the same level as new bricks. The bricklayers were satisfied building reused bricks, and there were no complaints about the small variations of the reused bricks

All stakeholders were satisfied with the results of the project. The building owner, Copenhagen City, collected the experiences and published them in Danish in 2017 (Copenhagen 2017).

5.4 MULTIPLE-RECYCLING PROJECT

In multiple-recycling projects, we deal with renewal of a greater urban harbour or industrial area. Many different types and sizes are presented in Chapter 2, Section 2.5.

Like single-recycling projects, the challenge of comprehensive integrated CDW management in multiple-recycling projects is the transformation of the CDW to demanded resources. The success of transformation of waste to resources depends on the deliverance of the sufficient amount of recycled materials with the right quality at the right place at the agreed upon time. The issue of matching recycled CDW with the demand of resources is one of the most important preconditions for recycling CDW. Crushing concrete and masonry, transportation and stockpiling and awaiting selling to a new construction project is not the optimal business case. The optimal business case is the demolition with a planned recycling and waste management process with a maximum of utilization and a minimum of logistic costs.

5.4.1 Planning and management of multiple-recycling projects

5.4.1.1 Initial considerations

The multiple-recycling project starts with the property owner/developers and their investors preparing a master plan for the development project. Often a due diligence of the whole area and structures, with respect to existing assets and environmental situation, forms the basis for the project. Depending on the size of the property and the requirements of the public authorities, the master plan includes a strategy of the development with short- and long-term plans, and the division of the property into plots. The master plan will also express the owner, developer and the investor's ambitions with respect to sustainability, circular economy, CO_2 reduction, the opportunities for certification according to LEED, DGNB, BREEAM or other certification schemes. In the perspective of a possible certification of the whole development project, or parts of it, the owner and his planners will take further action.

5.4.1.2 Assessment of the potential of reuse, recycling and utilization of existing structures and buildings

Referring to the purpose and master plan of the development, the existing buildings and structures are assessed with respect to their quality and application to the master plan. Based on this assessment, the individual buildings and structures are assessed with respect to their potential of reuse or recycling, as mentioned under the section above on planning and management of single recycling projects.

5.4.1.3 Assessment of opportunities for substituting natural resources and building with recycled materials in the new project

Besides reuse of buildings and structures, a development project needs a lot of resources for infrastructures, especially resources for temporary and permanent roads, RAC and structures, and so on, as mentioned under the single recycling project. Assessment of the medium and long-term needs and planning of the recycling processes should be part of the master plan.

5.4.1.4 Logistics and timing

The logistic planning and management is one of the most important and decisive issues of the single recycling project. This is also true for multiple recycling projects. To meet the goals of recycling, all processes of recycling materials and use of the materials should be carried out on-site. Recycling of materials requires space for handling and preparation of the materials for reuse/recycling. Therefore, a detailed logistic based on the master plan must be prepared. The plan should include the following:

- Plan of demolition of buildings, structures, and handling CDW, including potential reuse and recycling of materials
- Plan of supply of reused and recycled materials in the new housing projects
- Matching plan for on time delivery of reused/recycled materials on the right individual plots and building projects
- Planned sites for crushing materials, stockpiles of materials before and after crushing, sites for collection and sorting, treatment of CDW materials and products for new buildings

5.4.1.5 Ownership responsibility

The owners of the plots and contractors of respective buildings and plots should enter a mutual agreement on handling CDW, including reuse and recycling of CDW materials and the logistic planning. The owner must take action to prepare an overall CDW management plan to be agreed upon by all stakeholders in the development project.

5.4.1.6 Responsibilities of quality and purity of the recycled materials

It is generally expected that the owner of the development project will direct the responsibility of quality and purity of the recycled materials to the owners or contractors of the specific plots in accordance to the description of the single recycling projects.

5.4.2 Decision of the multi-recycling project

The decision to carry out the demolition project and the construction project on the basis of integrated CDW management or on the traditional basis without recycling materials, depends on a systematic analyis of the actual opportunities and above-mentioned considerations. The analysis can be based on a LCA approach, or on a simple assessment of the most important parameters, for instance: economy, risks, number of worksites, energy and CO_2 emission.

5.4.3 Implementation of the multiple-recycling project

The implementation of a multiple recycling project takes time; up to 25 years or more is not unusual. The owners and planners must be prepared for changes, especially changes of prices of products and other resources, which can make changes of the planned recycling necessary. However, delays, stressed economy and other changes must not cause a lack of attention to opportunities for prevention, reduction, reuse, recycling and recovery of CDW.

5.4.4 Case stories

5.4.4.1 Transformation of the Frederiksberg cable work, Nordic Cable Industry (NKT), Copenhagen

In 1985–1990, a major cable work was closed and the area developed for a new combined residential and university area, as shown in Figure 5.6. The demolition of the *NKT Cable Work* from 1985 to 1986 was a milestone in

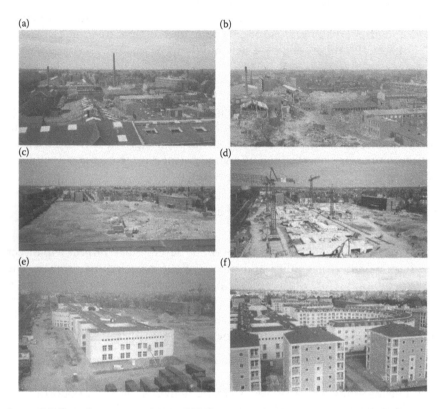

Figure 5.6 Transformation of old NKT Cable Works to new campus of Copenhagen Business School (CBS) 1985–1990, an example of multiple-recycling. The picture (a) shows the start of the NKT demolition work with blasting of the chimney, 1985. The ongoing demolition, 1986, is shown in picture (b). The crusher was installed in a major industry hall, on left hand in the picture, to protect the neighbours against dust. On right hand the stored brick and concrete prior to crushing is shown. Approximately 80,000 tonnes of demolition waste were generated. The picture (c) shows the site after demolition prepared for the construction of the new buildings. The start of the construction of new buildings is shown in picture (d) and picture (e) and (f) show the completion of the construction. The only reused building was the administration building, which is seen in the background, top right in the pictures, clearly in picture (d). Most of the rubble was crushed on-site and reused. The Danish architect Henning Larsen designed new buildings for the CBS, which were completed in 1990.

the recycling business in Denmark, because it was the first time crushing of CDW on-site and recycling of crushed materials was introduced. The area to be developed comprised approximately 6 hectares and 60,000 square meters of floorage area. Most of the buildings within the area were demolished. The administration building with a green roof – on the top right of the pictures (Figure 5.6), was rehabilitated and reused. Approximately 80,000 tonnes of demolition waste were generated. Most of the rubble was crushed on-site and reused. Figure 5.6b shows some stocks of rubble prior to crushing. The Danish architect Henning Larsen designed new buildings for the Copenhagen Business School in Figure 5.6e,f.

5.4.4.2 Development of the Marble Pier – The Port of Copenhagen

The development of the Copenhagen Harbour areas is typical of many similar harbour projects in other cities. The harbour development comprises all element of a multi-transformation scenario: preservation of old cultural heritage, reuse of silos and infrastructure and construction of new buildings for residential as well as cultural and commercial purpose.

The major part of the development began with the demolition of the former UNICEF warehouse and construction of the UN City Copenhagen, the regional headquarter of 11 UN agencies, situated on the Marble Pier. The demolition project comprised demolition of the UNICEF 30,000 m² storage building and the 3,000 m² administration facilities, producing approximately 50,000 t of CDW. The buildings were constructed in the 1980s and did not contain asbestos materials or other kind of hazardous materials. 90% of CDW was recycled. Furthermore, the old marble work at the end of the pier was also demolished. The new UN building was put into use in 2013, and the building was awarded Platinum, according to the LEED certification Figure 5.7

5.4.4.3 Development of Olympic Park – London

After selection of London for hosting the Olympic Games 2012, the 2.5 km² site in East London was planned to facilitate the games. The Olympic Delivery Authority (ODA) was responsible for its development and in 2007; ODA published a sustainable development strategy for CDW management. The demolition work comprised 215 buildings. The objectives and targets established in relation to CDW and resource efficiency were the following (Deloitte et al. 2012; Carris 2014):

Objectives

- Optimise the opportunities to design out waste, and to maximise the reuse and recycling of material arising during demolition, remediation and construction

2010

2014

Figure 5.7 Development of the Marble Pier, Copenhagen Harbour. Demolition of the old UNICEF warehouse, 2010; and construction of the new UN CITY Copenhagen, 2014. The old house, former office of the harbour pilots, has been preserved for cultural reasons.

- Operate within the ODA's waste hierarchy of: eliminate, reduce, reuse, recycle, recover, and dispose
- Minimize the amount of waste to be removed from the site and use sustainable transport methods where possible

Design out Waste (DoW)

- The ODA was seeking to design out, as far as practical, the production of waste during construction and operation of the facilities. When designing venues, materials selection and structure will be kept under review to help minimize waste
- At least 20% by value of materials to be from a reused or recycled source
- At least 25% recycled aggregate by weight for the permanent venues and associated park-wide infrastructure
- Across all of the elements of the park, care will be taken to optimize the provision of permanent elements for legacy, and temporary elements for game time
- To minimize any waste during the conversion from games to the legacy phase, all temporary venues and structures will be designed with reuse and recycling in mind

Table 5.6 CDV management targets and achievements – London 2012 Olympic Park

CDW Management targets	CDW Management achievements
• 90% reused or recycled demolition waste by weight • 90% reused or recycled construction waste by weight • 20% of materials to be from reused or recycled source by value • 25% recycled aggregate by weight	• 98.5% (427,531 t) reuse and recycling • 22% (approximately 170,000 t) aggregates from recycled sources • 20% t new material were saved • On average structural concrete contained 30% recycled aggregates • Over 20,000 lorry movements were saved

Source: Deloitte, Bio, BRE, RPS, ICEDD, VTT, and FCT, Resource efficient use of mixed wastes—Task 2-case study: Construction works in the preparation of the Olympics Games in London 2012.

Demolition and Remediation

- At least 90%, by weight, of the material arising through the demolition works will be reused or recycled
- The topographical modelling and demolition waste strategy should minimize the export of material and minimize the import of clean material or secondary aggregate

Construction

- At least 90%, by weight, of the material arising through construction to be reused, recycled or recovered
- Waste should be managed based on the proximity principle and source segregation

According to Deloitte et al. (2012), two of the most important lessons from the learning legacy were that many of the environmental sustainability benefits go hand in hand with cost savings, and that with the right approach to projects of this scale it is possible to drive innovation in areas such as design and materials specification, see Table 5.6. The ODA's overall objective of achieving exemplar sustainable waste management was achieved on a project of significant scale, with challenging and inflexible deadlines.

5.5 THE IRMA PROJECT

From 2003 to 2006, a project called *Integrated Decontaminated and Rehabilitation of Buildings, Structures and Materials in Urban Renewal (IRMA)* was conducted with the participation of 16 private and public organizations from 7 countries (Belgium, Denmark, Germany, The Netherlands, Norway, Spain and the United Kingdom). The EU supported the project with funds from the EU Commission Frame

Work 5.[1] The project members represent important stakeholders with interests in the decontamination of buildings in urban development: housing and civil contractors, demolition contractors, recycling specialists, consultants, universities, research institutes and municipal administrators. The main objective of the IRMA Project was to develop and implement a general *City Concept* including a toolbox of expert systems.

In the context of the IRMA Project, a *city* means an integrated building area under development with a number of structures, including infrastructure and supply installations. Typically, a city includes the following structures: a big industry complex, a harbour area, a big hospital, an airport, a railroad station, several housing quarters or a combination of the former. The City Concept addressed development projects of areas with the size of at least one hectare (10,000 m²), typically several hectares, where the demolition of old buildings and construction of new buildings take place over a long period.

5.5.1 The city concept

The IRMA *City Concept* model is a holistic way of thinking, and a specific way of doing with respect to the management of buildings and building materials in urban renewal. The model consists of:

- A description of the generic processes that constitute an urban development project
- Related tools for supporting some of these processes, including General Information System (GIS)

The generic *City Concept* process is the chain of processes that have to be executed from the start of an urban development project – named *Old City* – to the completion of the development project – named *New City* in accordance to the multi-building transformation scenario described in Section 5.4 and illustrated in Figure 5.2c. The basic philosophy of the concept is to organize and manage the metamorphosis of the *Old City* to the *New City* with respect to all processes concerning total and partial demolition, cleansing of polluted buildings and materials, recycling and logistics management of materials, as shown in Figure 5.8 and described in Table 5.7.

The decisions about the renewal strategy are based on many factors. The contamination of structures and materials are among the determinant factors. A sequence of measures to identify and manage contamination

---

[1] European Commission Fifth Framework Programme Energy, environment and sustainable development.
Key Action 4: City of Tomorrow and Cultural Heritage, Contract no. EVK4-CT-220-00092, Partners: NIRAS DEMEX, Dansk Betonteknik A/S, DEMEX A/S, Brandis A/S, SBS Byfornyelse, Meldgaard A/S (Denmark), Intron BV, Rotterdam Public Works Engineering Consultants, Delft Technical University (The Netherlands), Demoliciones Tecnicas S.A. (Spain), Belgian Research Institute, Enviro Challenge, Brussels Institute for Management of Environment (Belgium), Contento Trade Srl (Italy), Dr. Tech. Olav Olsen a.s. (Norway), Federal State of Bremen, Hochschule Bremen, Laboratory for Building Materials (Germany).

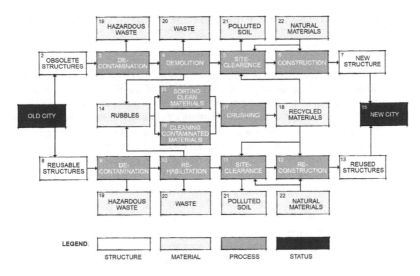

Figure 5.8 Principal sketch of elements of the urban transformation model. (From Integrated Decontamination and Rehabilitation of Buildings, Structures and materials in Urban Renewal [IRMA], 2006.)

will be designed within the first stages of the project. These measures are implemented into the integrated management system with the following consequences for the urban renewal process:

- Hazardous substances in buildings and structures are identified and classified
- Toxic emissions during demolition activities are prevented
- The lifetime of a structure can be extended
- Existing polluted building and structures can be rehabilitated and reused
- Important volumes of polluted demolition waste materials (primarily concrete and masonry rubble) can be recycled
- The volume of contaminated building waste will be reduced
- Materials can be recovered upon complete or selective demolition to save natural resources, and to avoid the need to dispose of waste
- Open spaces at the urban rehabilitation sites are used for the establishment of crushing and sorting machinery in order to allow all recycling to be carried out on-site. Only contaminated waste and other non-recyclable materials are to be transported from the site

A very important factor, probably the most important one, is the political factor. The success of an urban development project depends on political acceptance. The political drivers are providing housing and city development with a signature of green-change thinking and preservation of cultural heritage. Therefore, rehabilitations of old buildings are often important political issues.

Table 5.7 Description of elements of the urban transformation model (Figure 5.8)

Element no.	Description
1	The *Old City* comprises a number of structures. Based on the assessment of the structures and depending on the master plan of the city development and on the condition of the existing structures, infrastructure, etc., the structures are divided in two main categories: obsolete structures and reusable structures
2	The *obsolete structures* comprise all kind of structures. The structures are assessed with respect to demolition and waste management, including screening of hazardous materials and mapping resources. A general master plan for demolition contamination, recycling and waste management is prepared
3	The *decontamination* comprises asbestos removal, collection of hazardous waste, removal of polluted substances and structures, cleansing of surfaces. A part of the decontamination will take place during the demolition work
4	The *demolition* is performed as selective demolition in order to sort all fractions of building materials on the site. Contaminated structures are split into clean and contaminated materials in order to minimize the amount of contaminated materials.
5	The *site clearance* takes place after completion of the demolition. Polluted soil will be removed from the site, and replaced by natural materials or recycled materials. The site clearance depends on the master plan of the new structures
6	New structures are constructed on the cleared sites with the maximum use of recycled materials for aggregate in concrete, sub-base materials or fill. Recycled materials are also used in construction of temporary roads and structures
7	The *new structures* comprise new buildings, roads, free space and gardens
8	The *reusable structures*, including buildings, roads and various installations are assessed with respect to their function in the new city
9	The *decontamination* comprises asbestos removal, collection of hazardous waste, removal of polluted substances and structures, cleansing of surfaces. Part of the decontamination take place during the following rehabilitation work
10	The *rehabilitation* of the structure depends on the need for changes of the building and the building conditions. An important part of the rehabilitation work is removal of installations and partial demolition of the structure. A part of the decontamination takes place during the rehabilitation work
11	The *site clearance* around and inside the structure starts after completion of the partial demolition. Polluted soil is removed from the site, and replaced by natural materials or recycled materials
12	The *reconstruction* of the reused structures starts according to the public requirements of new structures and the design of the reused structures. Recycled materials are used to the widest extent
13	The *reused structures* comply with the public requirements for buildings and other structures, and the lifetime is expected to be the same as for new buildings
14	The *rubble* generated from demolition and rehabilitation are collected and transported to a building waste recycling plant on-site or transported to an external plant according to the demolition waste recycling plan. Clean materials and contaminated materials are stored separately
15	The *clean rubble materials* (not contaminated) are sorted for steel, wood, etc.

(Continued)

Table 5.7 (Continued) Description of elements of the urban transformation model (Figure 5.8)

Element no.	Description
16	The *rubble materials contaminated* by oil, chemicals, and heavy metals are cleaned or reused for special purposes
17	*Crushing* of rubble materials takes place in mobile crushing plants and sieves on-site or in stationary crushing plant outside the city area
18	The *recycled materials* are used as aggregates, sub-base materials or filling materials
19	The *hazardous waste* materials must be incinerated, disposed off at controlled landfills or subjected to special treatment
20	The *waste,* which is not recycled is disposed of at landfills or incinerated in accordance to national and local regulations
21	*Polluted soil* is to be removed according to national/local regulations and the intended future use of the site. The polluted soil might be treated and returned to the site
22	*Natural materials* or virgin materials are transported from quarries or riverbeds outside the City
23	The *New City* will contain a mix of new and reused structures. The infrastructure will also be new or reused structures

Note: Modified according to Integrated Decontamination and Rehabilitation of Buildings, Structures and materials in Urban Renewal (IRMA), 2006.

An objective of the *City Concept* is to optimize the material flow with respect to economy and environment. It means that the amount of generated waste materials and the consumption of natural materials must be reduced to a minimum, which requires maximum recycling. Reducing waste and substituting natural resources with recycled materials is a win–win situation, like the principle presented in Section 5.2 and Figure 5.1.

The contamination of buildings to be demolished or rehabilitated is a crucial barrier to recycling. Hazardous waste is very expensive to dispose of on controlled landfill, to incinerate or to receive special treatment. Cost reduction can be obtained by separating the contaminated materials from clean materials, in order to minimize the volume of contaminated materials. Additionally, costs can be reduced by the development and implementation of appropriate cleansing techniques for buildings, structures and materials (see Section 3.5).

In principle, the comprehensive and integrated approach does not deal with major urban development projects only, but also minor demolition and recycling projects. In the City Concept, we refer to multiple-building transformation, where demolition and recycling projects are integrated in the master plan of the urban development. In smaller common building projects, the demolition and waste handling project are often carried out as separate projects (single-recycling projects), often without specific opportunities for recycling in new building projects. The key issues of the transformation processes are summarized in Table 5.8.

Table 5.8 Key issues of the IRMA City Concept

Process	Key issues
Demolition Selective demolition will be performed in following stages: • Asbestos removal, clearance of hazardous materials • Stripping of the structure, including removal of doors, windows, roof and installations • Demolition of bearing structures, including beams, walls and plates • Clearance of the ground area	• To identify and assess the hazardous materials in existing buildings and structures • To clean surfaces and/or separate hazardous materials from clean materials as a specific process of selective demolition • To reduce impact to the environment and risk of occupational health problems during demolition of contaminated buildings • To save resources
Recycling Recycling focuses on: • Recycling of concrete and masonry • Reuse of bricks, stones and timber • Reuse of building products, doors, windows, etc.	• To control hazardous materials in recycled materials • To reduce the risk of impact to the environment and occupational health from hazardous materials in the recycled materials, both during the recycling processes and during lifetime of a structure built with recycled materials • To transform waste to resources
Decontamination of materials Decontamination of CDW materials to be recycled	• To develop economical feasible methods for cleaning polluted materials • To establish criteria for clean recyclable materials
Rehabilitation Rehabilitation of buildings and structures comprises: • Partial demolition • Reconstruction of structures • Repair and/or renewal of doors, windows, roofs and other installations	• To identify and assess the hazardous materials in existing buildings and structures • To clean surfaces and/or separate hazardous materials from clean materials • To reduce impact to the environment and risk of occupational health problems during rehabilitation of contaminated buildings • To extend the life cycle of buildings and structures
Reuse of buildings Reuse of: • Complete buildings • Parts of buildings • Infrastructure, e.g. roads and bridges	• To control hazardous materials in the reused structures • To reduce the risk of impact to the environment and occupational health from hazardous materials in the reused structures, both during the reconstruction processes and during the lifetime of a reused structure • To transform the life cycles of buildings

(Continued)

Table 5.8 (Continued) Key issues of the IRMA City Concept

Process	Key issues
Decontamination of buildings The decontamination process comprises: • Removal of major contaminated structures • Removal of contaminated surface layers until a certain thickness • Surface cleansing • Handling of contaminated materials	• To develop economically feasible methods for cleaning contaminated structures • To establish criteria for clean structures • To prepare transformation of buildings

Source: Integrated Decontamination and Rehabilitation of Buildings, Structures and materials in Urban Renewal (IRMA), 2006.

5.5.2 Demonstration projects

By the end of the IRMA project, the City Concept was demonstrated and evaluated on the basis of seven ongoing European projects, briefly presented in the summary of the IRMA project (IRMA 2006):

• Decontamination and demolition of an old gas production plant in Aarhus (Denmark)
• Strategies for decontamination and recycling of three buildings in Bremen, Germany
• Demolition of buildings creating space for the new NATO Headquarters in Evere, Brussels
• Pre-Demolition Assessment of buildings in Brussels
• City development of the Carlsberg Breweries, Copenhagen
• Demolition of the Kubota España factory, Madrid In 2006
• City development the Merwehaven Port Area of Rotterdam

5.5.2.1 Old gas production plant in Aarhus (Denmark)

The gas production plant has been in operation from 1912 to 1968 (owned by the municipality of Aarhus). The whole plant covers an area of 6,800 m². The area comprises five buildings and the concrete foundations of two gas tanks. The municipality intended to build dwellings for elderly people on the area.

The objectives of the Gasworks Demonstration Project were:

• To demonstrate the methods of assessment of contaminated buildings
• To demonstrate the decontamination technologies developed during the IRMA project

The results of the project indicated that contamination from the operation of the gas works had primarily an impact on the surfaces of the buildings. Samples taken from joints and bricks deeper inside the building parts showed only very modest overruns of the Danish EPA soil criteria. Based on the results of the investigations, it was estimated that out of a total of approximately 4,500 tons of building materials on the gas works site, only about 640 tons were certain to be contaminated. The potential for reuse of crushed building materials was substantial if the materials were cleansed before crushing and if the crushed material was to be used as backfill. It was estimated that recycling might reduce the costs by up to 90%.

5.5.2.2 Three buildings in Bremen, Germany

Three building in Bremen: a foundry building, a wool laundry and processing plant and a cold storage building. The objectives were to demonstrate and evaluate assessment and strategies for demolition, decontamination and recycling of the buildings.

The evaluation of the assessment process approach proposed by the IRMA guideline showed a good coverage of all essential steps of a qualified assessment. In addition, the workflow is depicted correctly as far as it is possible to describe the varying scenarios in real project work. In general, the proposed strategy goes far more into detail than the individual cases require, because the IRMA Guideline tries to cover a wide variety of scenarios.

5.5.2.3 New NATO Headquarters in Evere, Brussels

For the new HQ, a military base, containing over 50 buildings, roads and sports infrastructure had to be demolished. Estimated CDW was approximately 200,000 t. The new NATO HQ was completed and put into use by the end of 2017.

Given the urban environment, the size of the project and the possibilities for recycling, this project was a perfect example to implement the IRMA knowledge and City Concept. The aim was to maximise the economic benefit of the project by optimizing the amount of recycled materials in applications such as road foundations and concrete on the site itself, avoiding a lot of transport and primary resource use and minimising the environmental impact of the works.

In cooperation with the NATO project team, the objectives were:

- To develop promising recycling scenarios (on-site/off-site recycling and no recycling) to save money by assessing the qualities (also contamination) and quantities of recyclable materials, and identifying future applications of the secondary aggregates on the site

- To optimize the management and logistics of the site minimizing transport distances, environmental impact and hindrance to neighbours
- To elaborate a proposition for the tender document on recycling of materials and demolition

Results showed that the implementation of the IRMA City Concept principles on a large-scale building project work in practice. A clear economic benefit was demonstrated by the NATO project team, and the elaborated tender proposal was used to a high degree in the final tender document, implying that a recycling scenario, with full on-site recycling and application (mostly well-known applications like road sub foundations and foundations) was inserted into the tender document.

Important learning points from this demonstration project are:

- A convincing economic benefit was demonstrated by the project team, using clear and objective figures
- The cascade model used for matching qualities and quantities of recycled materials is a good way to provide several options
- Recycling on-site in high value applications (like structural concrete) needs convincing arguments like standards and practical examples

According to Janssen et al. (2012), the reduction of CO_2 emission and substitution of primary resources related to the three scenarios has been assessed. The assessment comprises the demolition for the new NATO headquarters, as well as the Rotterdam Harbour demonstration project presented below. The results are presented in Tables 5.9 and 5.10. The scenarios applied are:

- *Scenario 0*: No recycling and application of primary raw materials only
- *Scenario 1*: Off-site recycling and maximum use of recycled materials on-site
- *Scenario 2*: On-site recycling and maximum use of recycled materials on-site

Table 5.9 Relevant absolute CO_2 emission per scenario

Scenario	NATO HQ	Rotterdam Harbour
Scenario 0: No recycling	2,420 t	4,170 t
Scenario 1: Off-site recycling	2,068 t	2,281 t
Scenario 2: On-site recycling	1,725 t	1,027 t

Source: Janssen, G. m.t., Ho, H. M., Put, J. A. L., *Global assessment of urban renewal based on sustainable recycling strategies for construction and demolition waste. Progress of Recycling in the Built Environment.* Final report of the RILEM Technical Committee 217-PRE, 2012.

Note: Calculation is based on CO_2 emissions 880 g/km for lorries and 260 g/km for vans.

Table 5.10 Relevant absolute excavation of primary resources per scenario

Scenario	NATO HQ	Rotterdam Harbour
Scenario 0: no recycling	538,293 m³	199,635 m³
Scenario 1: off-site recycling	412,268 m³	69,137 m³
Scenario 2: on-site recycling	411,269 m³	65,489 m³

Source: Janssen, G. m.t., Ho, H. M., Put, J. A. L., *Global assessment of urban renewal based on sustainable recycling strategies for construction and demolition waste. Progress of Recycling in the Built Environment.* Final report of the RILEM Technical Committee 217-PRE, 2012.

The conclusions, according to (Janssen et al. 2012), on both urban renewal case studies indicate that a saving of up to 80% in CO_2 emissions can be realized by recycling on-site, compared to applying only primary raw materials. Moreover, a 70% decrease in the need for natural resources is feasible.

5.5.2.4 Pre-demolition assessment

The objective of the Pre-Demolition Assessment was to apply the knowledge on contamination assessment available from the IRMA project on actual case studies in Brussels and evaluate the practical tool for building assessment, primarily in the Brussels context, but also for wider applications. The assessment was done on two typical Brussels buildings: offices and dwellings.

It was concluded that the methodology and tools developed could also be successfully applied to smaller projects, with the following remarks:

- Within housing and regular working environments, no real 'special' contaminations are to be expected, so focus is best laid on the most important contaminants. Hence, with a view towards recycling of aggregates, the following contaminants were selected: PAHs, heavy metals, mineral oil and PCBs
- The tools developed within IRMA provide a good insight in possible contamination suspicions. However, the first source of information used within the case studies came from archives of the community, since those contained 'project-specific' information, whereas the IRMA knowledge is more generic and can serve as valuable background knowledge and supplementing information, thus as a 'checkup'
- The guideline for the developed building assessment was considered to be a valuable guidance document

5.5.2.5 The Carlsberg Breweries, Copenhagen

The Carlsberg Breweries intended to close down the production lines in Valby by the end of 2008. A great part of the 320,000 m² large area and a high number of the approximately 240 buildings located on the site, see Figure 5.9 – including approximately 200,000-m² of floor area – was considered for new urban development. In the beginning of 2006, Carlsberg took the first steps to initiate the planning process of the development. However, at that time, no details on the future use of the existing buildings, or the area, had been defined or agreed upon – so all opportunities were in principle open.

The objectives of the Carlsberg Demonstration Project were to develop a baseline study. The study comprised a description of the overall building mass, the estimated degree of contamination, and the total amount of material sorted by categories, and so on. Finally, a preliminary master plan was prepared describing the opportunities for reuse of the existing buildings, and/or the potential for recycling the demolished materials in order to support the planning.

Results showed that the amount of building materials in all buildings to be demolished was estimated at approximately 360,000 t. The total amount of all building and construction materials, including roads, pavements and gravel was estimated at approximately 500,000 t. Carlsberg decided to reuse a number of buildings, approximately 80,000 t, including the preserved buildings, the existing old power station and the boiler house, the gable of terraces to the garden, the old administration building and some other houses. 390,000 t were recyclable, of which 210,000 t could be recycled on-site. Only 30,000 t were classified for disposal.

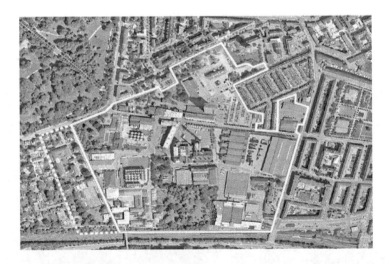

Figure 5.9 The Carlsberg site in Copenhagen. The line indicates the total area of the brewery being developed for the new Carlsberg City. The area inside the dash-dot line is not included in the development project. (From Integrated Decontamination and Rehabilitation of Buildings, Structures and materials in Urban Renewal [IRMA], 2006.)

A detailed study of the possible applications of recycled materials was conducted in 2005–2006. Three scenarios of recycling were examined. The main results of the assessment of the most idealistic recycling scenario showed that the total savings in one specific set up, including the value of gravel from existing roads, amounted to approximately DKK 50 million (EUR 6.7 million). However, in spite of Carlsberg's intention of recycling CDW, the implementation of the urban development of the *Carlsberg City* did not create the opportunities and preconditions for the idealistic multiple-recycling scenario.

After closing down in 2008, Carlsberg started the urban development. Until demolition or rehabilitation, the industrial buildings and facilities were reused for a great number of activities, especially of cultural character. In 2012, the first stage of the Carlsberg development began with the demolition of the bottling plant and preparation of the site for an education campus and a 100 m high tower, which is shown in Figure 5.10. The bottling plant building was decontaminated and demolished. Referring to the Irma project in 2006, the amount of the total building was assessed to 51,000 m² and 60,000 t of materials. Because of unforeseen mercury impurities, the recycling of concrete was restricted. The Danish Road Directorate accepted the concrete material to be recycled as crushed sub-base material in a highway north of Copenhagen.

In 2017, more than 75% of the buildings destined for demolition have been demolished. Because of lack of space and logistic problems, the proposed recycling of concrete on-site has not taken place. However, a

Figure 5.10 Demolition of the bottling plant, 2012–2013, and construction of the new buildings on the Carlsberg City site, 2013–2017.

lot of recycling of other CDW materials has taken place during the urban development until now. The rehabilitation of existing buildings worthy of preservation has called for reuse of façade bricks, and Carlsberg has developed very efficient methodologies and processes for selective demolition and cleansing of old bricks for reuse and repair of façades (see Figure 5.11).

The case story of development of the old Carlsberg Breweries into a new modern part of Copenhagen is an excellent example of a multi-recycling project with a mix of existing buildings to be rehabilitated, and buildings

Figure 5.11 Reuse of bricks from demolished building for repair of façades of buildings to be preserved.

to be demolished for new buildings. In spite of an estimated recycling potential of the magnitude of several hundred thousand tonnes only little recycling took place on-site. Because of construction activities and other activities, there was no space for the establishment of temporary stockyards for concrete or placing mobile concrete crushers. Therefore, all concrete materials were removed and recycled outside the Carlsberg site. It was not economically feasible to take back the crushed materials for recycling in the new buildings or structures.

5.5.2.6 Kubota España

Kubota España is a Japanese-owned company involved in the manufacturing of tractors. Kubota planned to close down their production and testing activities in the Madrid facilities by the end of 2006. The property covered approximately 75,000 m² consisting of 25,000 m² of building surface (14 independent building, 10 annexes) and approximately 50,000 m² of road/parking/test tracks. The Spanish IRMA partner DETECSA was awarded the contract for demolition as a turn-key project (i.e. excluding soil cleaning). DETECSA was in charge of the overall management of the demolition project that included total demolition of the buildings and structures, removal of pavement surfaces, removal of concrete elements down to 1 meter below surface, cleaning of current installations and handling CDW.

The objectives of the Kubota Demonstration Project was to assess the buildings for contaminants, estimate the degree of contamination, examine sampling procedures, assess general risks and develop health and safety procedures, develop a scheme for selection of subcontractors, describe the demolition process, define the potential for recycling and need for waste management.

The results obtained by the analysis of the costs of management of the residues of demolition obtained in Kubota for three different scenarios were:

- Transport + Landfill 21–22 €/t
- Transport + Recycled off-site 15 €/t
- Recycled on-site 8 €/t

Reduction of costs of recycled and waste management were approximately 48% in the case of recycling off-site. If recycling on-site, the costs could be reduced approximately by 58%. The cost of material execution would be reduced by approximately 45% if the materials were recycled on-site. The demolition material represents a valuable product instead of becoming a cost when managed like waste (to landfill).

In Kubota's project, about 70% of potential recycled materials would be removed. Finally, the use and development of management tools for building contamination was very interesting to the users of structures and refurbishment/demolition enterprises.

5.5.2.7 Merwehaven Port Area of Rotterdam

In 2006, a part of the Merwehaven Port Area of Rotterdam was planned to be transformed into housing and leisure. A study was carried out to investigate the possibilities of reusing the stony fraction of the CDW from the area, and reuse or recycle it in the estimated 4,000 houses that were thought of in the preliminary studies of City Development. These houses would have to have been built by 2015. The study was mainly carried out by using digitally available data, GIS and computer models, since urban planning is done more and more using GIS. A total quantity of approximately 225,000 tons of concrete, masonry, pavement materials and aggregates were available for recycling (conservative estimate). Potentially there is more, in the form of quay constructions and buried foundations. There are also large volumes of metals and installations available, which could either be reused or recycled but were not part of the inventory.

The City Ports project area (1,400 ha) is situated in the port area of Rotterdam. Approximately 20,000 people are employed in the area and it houses 2,000 inhabitants. The project area was largely a polder in the nineteenth century, until harbours were dug and industries were built and it became a beehive of very diverse economic activities. In the course of time, some buildings were demolished, while others were built. In the future, harbour-related activities will remain, but other economic activities will be developed, and in time houses will be built. Different aspects are included in the spatial planning, including such diverse items as: energy storage, ecology, contamination, sub-surface obstacles (archaeological remnants, unexploded objects, cables and pipes, remnants of foundations), subway tunnels, the relationship with structures above ground, air quality, noise, flood risks, and so on. City Ports Rotterdam has incorporated the ideas and opinions of local organizations, businesses, pressure groups and residents into the Development Strategy.

The main objectives of the City Ports Rotterdam demonstration project were:

- To develop a baseline study describing the overall building mass and its composition and possible contamination
- To develop a preliminary master plan describing the opportunities/ strategies for the reuse of the existing buildings or the recycling of materials if the buildings are to be demolished in the light of future developments

Four scenarios comprising off-site and on-site recycling were studied using expert systems with calculation of logistic parameters. The outcome was as follows:

- Time and space constraints were arbitrarily set to 2 years and 6 hectares, respectively. It turned out that activities could be done quicker on less space. Crushing on-site is a noisy activity, but proved to remain within acceptable levels for any vulnerable object like houses, schools or offices nearby. In practice, this means that logistic improvements can be made or, that problems arising in the process may not be critical
- The environmental issues of the scenarios differ substantially. The Expert System calculated the logistics of the scenarios
- Total road transport distance could be reduced by two-thirds, from 6.3 million to 1.5 million kilometres. Total CO_2 production could be reduced from 5,560 tons to 1,440 tons, a 74% reduction
- The differences are of such magnitude that one may expect they outweigh differences in environmental performance of stationary vs. mobile crushers and concrete plants
- A rough calculation was made of the relative cost of the different scenarios. Crushing on-site and reusing the aggregates on-site proved cheaper by an order of magnitude of €3,000,000.00. Concrete production on-site was not cheaper. Off-site crushing was more expensive than on-site crushing

According to Janssen et al. (2012), the CO_2 emission and potential substitution of primary resources for three scenarios have been assessed (see Tables 5.9 and 5.10).

Chapter 6

Disaster waste management

Next to saving lives, medical care, sheltering and feeding people, the handling of debris is the most challenging problem in the emergency response.

Lessons learnt from a number of conflicts and disasters during the past 30 years

6.1 INTRODUCTION

One of the greatest technological challenges of our time is to prevent and relieve damages to cities, and to protect society from causes of natural disasters and conflicts. This challenge includes the management of waste materials from the structural damages, disaster waste management (DWM), which in principle resembles CDW management discussed in the previous chapters of this book. Like integrated CDW management, we are also talking about integrated DWM focusing on recycling and the use of recycled resources, especially debris, in reconstruction of damaged buildings and infrastructure.

In late 1988, UNESCO and the Secretariat of RILEM[1] discussed the possibilities of cooperation concerning earthquake disaster relief in the light of the earthquake in Armenia, in 1988. In order to meet the requirements of UNESCO, the RILEM Technical Committee on Demolition and Recycling (TC-121-DRG) decided to deal with demolition and recycling after disasters as one of its main objectives. The work resulted in the RILEM report *Disaster Planning, Structural Assessment and Recycling* (De Pauw and Lauritzen 1994). Based on European and American experiences, it was concluded that the present technical level of know-how in the field of recycling and reuse of building and construction materials was sufficient to implement such techniques into the continuous rehabilitation process which

---

[1] Réunion Internationale des Laboratoires d'Essais et de Recherches sur les Matériaux et les Constructions / International Union on Testing and Research Laboratories for Materials and Structures.

takes place in urban areas all over the world after disasters and wars. In addition, the experiences from reconstruction of Europe after World War II were considered.

The demand for DWM is global, and this demand is much more comprehensive and long-term oriented than the demand for emergency humanitarian relief. There is a great need for assistance, especially in the form of preventive and mitigative measures, the demand for mitigative measures being short-term by nature. In order to give effective disaster assistance, it is necessary to look at the assistance as an integrated entirety – Integrated Disaster Management and DWM – based on cooperation across professional and political barriers. In the UN and The World Bank's policies regarding disaster assistance, emphasis is placed on preventive arrangements and analysis of the 'anatomy' of the disaster and its consequences. It is, as an example, important as soon as possible in the disaster development to aim at a reconstruction of the infrastructure and supply lines of the stricken area. Recycling of building materials from destroyed structures will in this case play an important role.

Generally, the challenges of integrated CDW management, as described in the previous chapters of this book, focus on handling and recycling of current streams of CDW from construction, repair and demolition works. The CDW management in most countries is based on a set of legal, technical and administrative instruments aiming on meeting the management goals and challenges of predicted and continuous CDW sources and streams. Disasters and conflicts generate unpredicted amounts of waste materials from destroyed and damaged buildings and structures, often within a very short period. For instance, earthquakes, tsunamis and hurricanes produce millions of tonnes of waste within a few hours, which need an immediate response.

Buildings and structures contain more than 90% recyclable materials, which can be utilised in shelters and new buildings/structures in emergency response and reconstruction, rather than being disposed of. Natural stones, roofing materials (such as steel plates and tiles), windows, doors and timber can be reused without special treatment. Plain and reinforced concrete, masonry bricks and asphalt, which represent the majority of building waste materials, need to be recovered and crushed (recycled) before utilisation. However, many buildings and structures, especially in urban communities, contain substances potentially hazardous to the environment and human health, hindering the reuse of materials. Substances considered harmful today, such as asbestos, polychlorinated biphenyls (PCBs), heavy metals, certain paints, and so on, may be present, which can make the clearance and removal of building waste materials problematic (see Chapter 2, Section 2.6). Debris management after conflicts and disaster should follow the general principles of CDW management. However, in case of an emergency, the necessary resources and capacity are not available and many recyclable materials are lost.

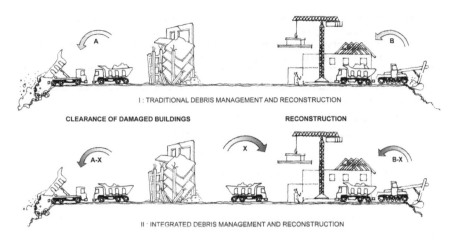

I : TRADITIONAL DEBRIS MANAGEMENT AND RECONSTRUCTION

CLEARANCE OF DAMAGED BUILDINGS   RECONSTRUCTION

II : INTEGRATED DEBRIS MANAGEMENT AND RECONSTRUCTION

Figure 6.1 Principle of business model of integrated DWM. Top: Shows a clearance work and a reconstruction project without any synergy. Bottom: Shows exploitation of the recycled debris as resources for the reconstruction of buildings. (See Table 6.1.)

This chapter will introduce debris management as a cross-cutting issue, giving an overview of debris (or rubble)[2] management experiences from earlier disasters and conflicts (De Pauw and Lauritzen 1994; Lauritzen 1994). Recommended strategies and actions for sustainable debris management and integration in post-disaster/conflict management are presented. The chapter includes actual case stories from the latest conflicts and natural disasters, such as Kosovo, Lebanon, Haiti and Nepal. Finally, a summary of experiences applicable for the reconstruction of Syria after the civil war are considered.

Referring to the economic model for integrated CDW resource management presented in Chapter 5, Section 5.1 and Table 5.1, a similar business model for integrated DWM and the assessment of the debris resource potential in a disaster scenario are presented in Figure 6.1 and Table 6.1.

The assessment comprises the following cost:

- Demolition and handling the disaster waste, focusing on debris handling
- Transport of debris
- Disposal of materials
- Recycling processes and production of reusable materials
- Cost of production of natural materials
- Transport of natural materials for reconstruction

---

[2] Solid waste from destroyed buildings and infrastructure are called debris or rubble. There is no principle difference in the terms other than the French and English wording.

Table 6.1 Principle of business model of integrated debris waste management illustrated in Figure 6.1. Assessment and comparison of total cost and environmental loads of traditional debris and reconstruction management and integrated debris and resource management

Type of debris management	Cost/CO$_2$
I. Traditional debris removal and reconstruction:	
Removal and disposal of A tonnes of debris	I.A
Providence of B tonnes raw materials for reconstruction	I.B
Total Cost/Load	I.A + I.B
II. Integrated debris removal, recycling and reconstruction	
Removal and disposal of A – X tonnes debris	II.A – X
Providence of B – X tonnes raw materials for reconstruction	II.A – X
Transformation, crushing, qualification of X tonnes recycled materials	X
Total Cost/Load	(II.A – X) + (II.B – X) + X

Preferences:

I.A + I.B > (II.A – X) + (II.B – X) + X → Integrated debris management to be preferred

I.A + I.B < (II.A – X) + (II.B – X) + X → Traditional debris management to be preferred

Besides the mentioned costs, the assessment must also consider the application of recycling debris to the DWM strategy and the strategy of reconstruction and other relevant issues.

6.2 DEBRIS REMOVAL – A CROSS-CUTTING ISSUE

The key issues of the experiences and lessons learnt from a number of conflicts and disasters during the past 30 years are the following:

- Next to saving lives, medical care, sheltering and feeding people, the handling of debris is the most challenging problem in the emergency response and early recovery processes because damaged buildings and debris hamper rescue work and block traffic
- Debris has a major impact on most of the UN clusters (see Figure 6.2) and there is a lack of organization and coordination of debris management related to the individual clusters
- The local authorities and the individual organizations – governmental organizations, as well as non-governmental organizations (NGOs) – involved in the emergency activities commence the movement and disposal of debris as they find it necessary without considering other clusters' interests

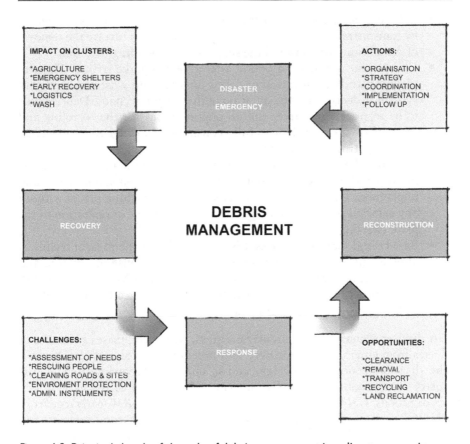

Figure 6.2 Principal sketch of the role of debris management in a disaster scenario.

- Property rights of building and landowners together with lack of administrative instruments are barriers to effective and expedient removal and management of debris and other kinds of solid waste
- Solid waste management including debris management is a cross cutting issue impacting the daily life of societies exposed to disasters, which requires immediate coordinated action and a specific strategy as early as possible during the disaster response

Besides the key issues of experiences and lessons learnt, the following should be noted:

- Generally, the importance of waste management after disasters has always been ignored or underestimated. Many problems could have been avoided if the authorities and others involved in pre and post-disaster planning and management had had a better understanding of opportunities and potential difficulties concerned with demolition and recycling

- In larger post-disaster reconstruction projects, the economy is dominated by transportation costs. These transportation costs involve the removal of demolition products and the supply of new building materials
- Chaos is inevitable for a period after an earthquake strikes. The pre-planning and organization of debris removal and waste management systems is decisive for the duration of the period of chaos. Debris needs to be seen as a resource, and all opportunities for the recycling and reuse of waste must be carefully considered in order to save energy, materials, time and money
- Recycling has been introduced after most of the major earthquakes. However, the recycling initiatives have always been spontaneous and unplanned. Donors have typically sponsored crushing plants without any detailed plan for the preparation of transport, or operation of the plant (see Figure 6.4)
- Based on global experiences it is concluded that the present technical level of know-how in the field of recycling and reuse of building and construction materials is applicable to the DWM in all countries and regions. Therefore, information, education and training of the stakeholders are very important

A disaster scenario has a circular progress, which comprises a number of phases from emergency to recovery and preparedness for a future disaster. The disaster responses are organized in clusters. According to UN, the cluster system comprises twelve clusters under control of the various UN organizations: UN OCHA (emergency coordination), UNDP (early recovery), UNHCR (protection of people), WHO (health), WFP (food security and agriculture), UNICEF (education) and others. Figure 6.2 presents an example of a simplified disaster scenario, which is explained below.

BOX 6.1　LIST OF MAJOR DISASTERS AND BRIEF DESCRIPTION OF EXPERIENCES OF DEBRIS MANAGEMENT AND HANDLING

The *El Asnam earthquake on 10 October 1980, Algeria* destroyed 4,000 buildings and generated approximately 3 million tonnes of debris. Belgium sponsored a pilot recycling project for the production of building blocks of concrete using recycled aggregates with successful results.

The German Red Cross sponsored a crushing plant in Armenia after the *Leninakan earthquake in December 1988*. The transport of the crushing plant was complicated. After installation, the plant produced materials for road construction.

After the *Loma Prieta earthquake on 17 October 1989, San Francisco*, several kilometres of double-deck highways were destroyed, and the reconstruction and retrofitting work took several years. Both demolition and waste disposal were problematic owing to the mix of waste materials, because organic materials were not accepted in the marine landfills in Oakland Bay (see Figure 6.3).

The extensive damage caused by the *Luzon earthquake on 15 July 1990, Philippines*, resulted in vast amounts of building waste, and thousands of people were made homeless. After the emergency rescue and demolition work, squatters moved to the sites of the destroyed houses and earned money on hazardous demolition work and recovering the reusable waste, such as steel from reinforcement bars.

After the *Erzincan earthquake in Turkey on 13 March 1992*, 6,500 dwellings were destroyed or severely damaged. Demolition waste, estimated at 500,000 to one million tonnes, was disposed of along roads and at improvised dumpsites outside the city, disturbing the traffic and polluting the environment.

In one day in *August 1992, Hurricane Andrew* produced 33 million m^3 of waste and damaged or destroyed more than 100,000 homes on the southern tip of Florida.

Following the *Northridge earthquake on 17 January 1994, Los Angeles*, 5.3 million m^3 of earthquake debris were collected, and either disposed of or recycled, according to City of Los Angeles estimates.

The *Great Hanshin-Awaji earthquake in Japan, on 17 January 1995* caused destruction of more than 100,000 buildings as well as infrastructure such as railways, roads and harbour facilities. Approximately 20 million tonnes of waste were generated, causing one of the gravest problems of the emergency relief work. Most of the debris was disposed of in Osaka Bay for land reclamation.

In *Lebanon, 17 years of war* have resulted in a 25-hectare area of uncontrolled disposal of all kinds of mixed waste, some of which is constantly burning due to self-combustion. Demolition and clearance of damaged buildings in the Central District of Beirut requires removal of 2–4 million tons of demolition waste. In 1994, demolition and clearance of the waste began. After sorting, the debris was crushed and used as land reclamation for the new part of BCD, *Normandie*.

The *Marmara earthquake in Turkey in August 1999*, generated more than 5 million tonnes of waste. Opportunities for recycling were missed because of a lack of understanding of the need for sorting the materials before crushing, and because of constraints between the Ministry of Public Works and the Ministry of Environment (see Figure 6.4).

The *Tsunami on 26 December 2004* caused an unknown amount – possibly more than 3 million m^3 waste – in Indonesia, Thailand and Sri Lanka.

Hurricane Katrina on 29 August 2005 generated 76 million m^3 (22 million tonnes) of waste, including a considerable amount of hazardous waste.

The *October 2005 Kashmir earthquake in Pakistan* caused 2.3 million m³ of waste, where 20% of steel and wood was recycled. Debris were dumped. Recycling was introduced supported by a crushing plant from Belgian donors.

Following the military attack on the *Nahr el-Bared Camp, Tripoli, Lebanon, 2007*, 20 hectares of densely populated land was destroyed and approximately 600,000 tonnes of debris contaminated by hazardous explosive waste, mines, booby traps, ammunition, air bombs, and so on, was produced. Most of the debris was recycled and used for coastal land reclamation, road construction and sub-base material.

The *12 January 2010 Haiti earthquake* damaged more than 400,000 buildings and the amount of rubble generated was estimated at 10 million m³. The rubble removal cost was estimated at $400USD million. The amount removed 18 months after the earthquake was 1.15 million m³.

The *Fukushima earthquake/tsunami, Japan, 11 March 2011* generated, according to the official figures, approximately 25 million m³ of mixed waste, including radioactive waste and waste taken by the sea. Approximately 400,000 buildings were destroyed or seriously damaged.

Hurricane Sandy, USA, 29–30 October 2012 caused total damages estimated at $75USD billion with more than 100,000 houses destroyed or seriously damaged. No estimate of the amount of waste has been available.

Nepal earthquake 25 April and 12 May 2015 caused damage to buildings temples in Kathmandu and Bhaktapur. Major damages occurred to rural areas around the two epicentres. The estimated amount of damaged buildings and building waste was approximately 10 million t.

Source: Updated list based on Lauritzen 1996/1997.

6.2.1 Impact on clusters

Disasters (typically earthquakes, hurricanes, floods and tsunamis) and man-made disasters (typically wars) generate vast amounts of waste within a very short time, especially building waste from damaged houses. Box 6.1 presents the amounts of debris from selected disasters. With reference to the experiences from the listed disasters and conflicts, typical impacts on some clusters of the UN cluster system are:

- *Agriculture*: Because of the urgent need for the removal of debris and rescue work, farmlands are blocked and spoiled because of the controlled and uncontrolled disposal of debris, often for many years. Figure 6.4 show an example from the Marmara earthquake in Turkey, August 1999

Figure 6.3 The Loma Prieta earthquake on 17 October 1989, San Francisco, caused destruction of several kilometres of double-deck highways. The reconstruction and retrofitting work took several years. Both demolition and waste disposal were problematic owing to the mix of waste materials, because wooden forms were found in the hollow boxes of the bridge and organic materials were not accepted in the marine landfills in Oakland Bay.

- *Emergency shelters and camps*: Following the damage to buildings and risks of future damage, there is an urgent need for emergency shelters and camps, which require suitable sites. In urban areas and densely rural areas, debris are disposed of at sites otherwise suitable for emergency shelters. Areas of destroyed buildings need to be cleared to make space for emergency shelters and camps. This was a serious problem in Haiti after the earthquake in January 2010 because of the lack of space. The flooding of lower areas that followed during the spring season reduced a suitable space for shelters and camps, which aggravated the situation
- *Early recovery*: Obstacles to early recovery activities caused by debris from damaged buildings and structures must be managed in a structured way. Immediate removal of debris for clearing roads and access to damaged buildings for the rescue of trapped persons should be carried out with respect to the debris management activities after the emergency phase
- *Logistics*: Debris blocks roads and hampers traffic. In order to provide food and emergency assistance, the roads have to be cleared as soon as possible. The removal of debris itself requires significant transport capacity

Figure 6.4 Deposit of waste after the August 1999 Marmara earthquake in Turkey. The figure shows a typical 'pancaked' building. An undamaged building of same type is seen in the background. The unplanned disposal of debris occupies valuable agricultural land. A crushing plant was donated to crush the debris to recyclable materials. However, it did not work because of lack of understanding of the use of the plant and the need for sorting debris from other kinds of waste before feeding the crusher.

- *Water sanitation and hygiene (WASH)*: All kinds of waste, especially household waste and human waste can cause epidemics. Debris in the streets and elsewhere attracts uncontrolled waste disposal and is a threat to WASH. Uncontrolled waste disposal is an accumulative problem. Therefore, debris should be removed from the streets and collected in controlled depots as soon as possible

6.2.2 Challenges

From international DWM experiences, it is well known that the handling of vast quantities of debris is hampered by a poor and often very late start of the planning process, lack of space for depositing the waste and a lack of instruments and procedures for permits to access private buildings. Furthermore, the provision of resources and mechanical assets for handling debris does not meet actual needs. The management of the demolition work, logistics and disposal activities are complicated and usually underestimated. Finally, everybody finds the issue of recycling debris and other building materials very attractive, but its implementation, taking into account economic and environmental feasibility is very poorly understood.

Some of the most important immediate challenges of debris management are:

- *Assessment of needs*: The overall problems and needs related to the debris and the emergency must be assessed as soon as possible. The assessment of debris follows the damage assessment, which often takes a long time. Many types of disasters, e.g. hurricanes and tsunamis, generate debris mixed with industrial waste, household waste, organic waste, etc. Therefore, it is important to provide an initial overview of the scope of problems related to the debris and needed resources for urgent actions
- *Rescuing people*: The overall need is to search and rescue people, which require resources and skills. Especially removal of collapsed parts of concrete buildings and entrance to unsafe structures must be carried out very carefully
- *Clearing roads and sites*: In order to get access to the rescue sites and distressed people for the provision of food, medical aid and other necessities, roads and sites have to be cleared. A prioritized plan for the clearing work must be prepared and the necessary resources must be provided. It is important the plan include collection or disposal sites for the debris and other types of waste. Special attention must be paid to handling dead bodies and human remains in the ruins and debris
- *Environment protection*: The handling of solid waste must take necessary consideration to environmental protection comprising special care of hazardous waste and prevention of drinking water. It is important to meet the accumulated need for disposal of household waste

- *Administrative instruments*: The provision of administrative instruments is an overall precondition for the emergency recovery and reconstruction, including debris clearing and management. In case of a disaster in a developed/industrial region, it is anticipated that the local authorities provide the administrative instruments and supervise their execution. In case of a disaster in poor/developing region without administrative capacity, external support for providing the administrative instruments is required by the UN. This was the case in Haiti after the January 2010 earthquake, where the city centre of the capital Port-au-Prince was impacted and many government offices were destroyed

6.2.3 Opportunities

In order to meet the challenges of the debris management, the opportunities must be assessed and evaluated as soon as possible after the disaster strikes. Many disaster-exposed countries and regions have prepared contingency plans for the response of a future emergency. Some plans include debris management plans including allocation of resources for clearance work, and designation of sites for disposal of debris. The Nepal government had prepared a strategic plan (Government of Nepal and IOM 2014) for demolition of damaged structures and management of the debris and other solid waste. However, because of stress during the emergency, the strategic plan was not executed and the designated sites were not used.

The temporary removal and disposal of debris requires space, and sheltering homeless people requires space. In the case of very dense populations, such as Port-au-Prince, Haiti, the open space for sheltering homeless people is a serious bottleneck. The need for cleared sites must be coordinated and prioritized very early in order to avoid conflicts of various interests. Therefore, in the case of Haiti, the clearance of debris was integrated into the Shelter Cluster during the first months after the earthquake. The space for shelters needs a proper surface/base and protection against rain, floods, hurricanes, after-quakes and other 'secondary' incidents.

- *Clearance*: All resources for clearance of debris must be prioritized and organized. Clearance of main roads are important to facilitate the human emergency response
- *Removal*: After clearance, solid waste and debris are removed and transported to designated disposal sites or collected to selected pick-up points. Clearance and removal should include sorting of the solid waste in recyclable fractions (e.g. concrete, scrap metal, bricks, tiles, timber and asphalt) and disposable fractions (e.g. plastic, paper, household waste and hazardous waste [asbestos, batteries, polluted materials, etc.])

- *Transport*: Transportation of debris and other solid waste is expensive and time consuming. Transportation needs careful planning and agreed sites for tipping off. There is a big risk of fly-tipping, which means that the driver dumps the waste on an unexpedient site
- *Recycling*: Already during the early emergency phase, the solid waste management should aim at recycling of all valuable materials, especially debris, bricks, timber and construction steel. However, because of the need for quick removal of solid waste during the emergency phase, the recycling issue must wait until the start of reconstruction
- *Land reclamation*: Finding solid waste disposal sites is a crucial problem in all disaster scenarios. Therefore, opportunities for land reclamation should be exploited. After the great Hanshin-Awaji earthquake of 17 January 1995 in Japan, most of the 20 million tonnes of solid waste materials were tipped off in Osaka Bay for land reclamation after proper sorting and removal of hazardous waste. The opportunities for land reclamation at the coast line of Port-au-Prince after the January 2010 earthquake were not exploited because of environmental considerations

6.2.4 Actions

Actions to provide materials, resources and capacity for the early emergency aid will start immediately after registration of the disaster. Depending on the region or country's policies, appeal for international support will be addressed to the UN, or the international aid will be declared unwanted. The latter was the case after the flood in Myanmar 2015. The actions leading to a successful implementation of debris management during the early emergency are followed by a transition to reconstruction and a normal situation depending on the following issues:

- *Organization*: In case of a major disaster, the UN Office for the Coordination of Humanitarian Affairs (UN OCHA) primarily takes action supported by national and international organizations. The organization of actions related to debris management during the early emergency, recovery and reconstruction depends on the donors and local capacity
- *Strategy*: Based on assessment of needs and damages, a strategy for the early debris management is prepared. Usually, countries and regions exposed to natural disasters have prepared contingency plans with strategies for various response actions. However, strategies for debris management and recycling of materials are not seen very often. As mentioned under the section above on opportunities, the government of Nepal had a debris management strategy, but it did not work after the earthquakes in 2016

- *Coordination*: Referring to the UN cluster policy, a cluster organization is set up to meet the challenges. Often debris management and management of shelters form a specific cluster
- *Implementation*: National and international organizations – governmental and non-governmental organizations – take care of the specific actions from early response to final reconstruction
- *Follow-up*: The actions during the emergency response cycle must be evaluated and followed up with contingency plans, strategies, etc., to meet the future disasters

6.3 GUIDELINES FOR DWM

Guidelines for DWM are found in

- DWM Guidelines, published in 2011 by the United Nations Office for the Coordination of Humanitarian Affairs and the United Nation Environment Programme (UNEP) supported by the Swedish Contingencies Agency (MSB) (UNEP/OCIIA 2011)
- Guidance Note Debris Management, published by the United Nations Development Programme, 2013 (UNDP 2013)
- US EPA: Planning for Natural Disaster Debris, March 2008
- US FEMA-325: Public Assistance Debris Management Guide, July 2007

The guidelines comprise demolition of damaged buildings and structures and management of debris including crushing of debris. Unfortunately, the guidelines do not provide any details on the recycling processes and use recycled debris.

The UNEP/OCHA Guidelines are in four parts:

1. Introduction and overview
2. General guidelines divided into the Immediate Response Phase, Early Recovery Phase and Recovery Phase. The latter includes information on Contingency Planning
3. Key considerations that are important throughout the process – for example, health and safety, stakeholder management and common risks by disaster waste hazard type
4. Tools and checklists implementing the guidance provided. These are in attached annexes

Phase 1 (Emergency Phase, divided into 0–72 hours, immediate actions and 72 hours onwards): Short-term actions focus on identifying waste streams and appropriate disposal of waste. Main streets are to be cleared to provide access for search and rescue efforts and relief provisions.

Phase 2 (Early Recovery Phase): The work focuses on detailed planning of waste management and operations, including logistic planning and management, separation of waste fractions and selection of disposal sites.

Phase 3 (Recovery): Includes implementation of the DWM projects designed in Phase 2, and continued monitoring and evaluation of the disaster waste situation.

The *Recovery Phase* is followed by planning of a DWM Contingency Plan for options to meet the challenges of future disasters. The Guidelines present a list of suggested contents for a DWM Contingency Plan.

The UNDP Guidance Note is based on UNDP operations in a number of countries and regions, especially Haiti and Lebanon. The Guidance Note starts with the conceptual framework focusing on the following principles, which are referred in more detail throughout the Guidance Note:

1. Maximising the engagement and skills development of local people and communities
2. Empowerment of women and promotion of gender equality
3. Building capacity and social capital through effective coordination, communication and partnerships
4. Environmental protection
5. Conflict sensitivity
6. Investing in disaster risk reduction (DRR)

The UNDP Guidance Note (UNDP 2013) describes the planning phase, developing the project document including the situation analysis and needs assessment, programme strategy, results framework, identifying and mitigating risks, management arrangements, operational support, partnerships, monitoring and evaluation, communication strategy and resource mobilization. Like the UNEP/OCHA Guidelines (UNEP/OCHA 2011), the planning part of the UNDP Guidance Note contains a number of useful tables and checklists.

Following the planning phase, the UNDP guide presents the project implementation with a number of key actions, tables and checklists. The section ends with a quick summary of some of the results achieved through implementation of past debris management projects. Tables and checklists are embedded in the text. Whereas the UNEP/OCHA Guidelines are general and apparently not based on specific disaster waste scenarios, the UNDP Guidance Note refers to specific experiences, which makes it difficult to compare the two papers. However, the papers together make a very good complementary basis for an approach to understanding and performing the DWM processes.

The objective of both guidance papers is managing the disaster waste as waste. Managing the disaster waste as resources is not the scope. Both papers consider that recycling is an important part of DWM, and they describe DWM processes aiming at recycling materials. However, recycling of waste materials is not among the key considerations in the UNEP/OCHA Guidelines, and the recycling and resource management is not mentioned in the list of principles in the UNDP Guidance Note, as mentioned above.

UN *Sustainable Development Goals (SDG)*, specifically Goal #12 *Ensure sustainable consumption and production patterns*, including Target #12.5 *by 2030, substantially reduce waste generation through prevention, reduction, recycling and reuse* states the need for resource management in DWM. Referring to the SDG, the need for CO_2 reduction and concepts of circular economy are explained in Chapter 1. DWM must include recycling aimed at the use of recycled materials during the reconstruction of destroyed and damaged buildings and infrastructures after a disaster.

6.4 STRATEGY FOR INTEGRATED DISASTER WASTE MANAGEMENT

The most important principles for debris management are:

- Organization and planning of the debris management, and activities focusing on early emergency, initial needs assessment and short-term rapid debris handling activities. The following activities of recovery, reconstruction and sustainable development – mid-term and long-term actions – must also be considered.
- Overall strategy for sustainable debris management, including strategy for pre-disaster planning of debris management. The initial strategy should focus on the designation of a number of debris management sites for recycling and disposal of debris and building waste materials based on logistic and economic optimization.
- Coordination of resource management interests at various levels – national, regional and community – and interests of various stakeholders – public authorities, private house owners, emergency relief organizations, GO and NGOs and others. The coordination of the interest of the individual clusters with respect to debris management should also take place.
- Implementation of rapid debris removal activities based on the planning and strategy and allocated funding provided through international donor channels. The implementation also comprises the provision of maps, land registers, administrative legal instruments for demolition and clearance of destroyed buildings – both privately and publicly owned.
- Follow-up activities, including monitoring, quality control, adjustment of planning and time schedule, pre-disaster planning of debris management, etc.

Based on the experiences from case stories and examples of DWM and other disasters mentioned in Box 6.1, and case stories presented in Section 6.6, Box 6.2 presents a proposed strategy for integrated DWM.

BOX 6.2   PROPOSED STRATEGY FOR INTEGRATED DWM

STRATEGY

Emergency actions:

1. To organize the waste management as soon as possible during the early emergency and to concentrate efforts on removal of waste for rescuing people and reduce risk of infection and other health cases
2. To control the waste streams and disposal of waste to designated areas sites

Recovery actions:

3. To recover debris and suitable fractions of materials with respect to recycling and to sort out hazardous waste for special treatment
4. To recycle debris and other materials for substituting natural resources in reconstruction works

Reconstruction actions:

5. To prepare reconstruction of buildings and structures with secondary resources
6. To put people into work based on cash for work/production
7. To minimize transport
8. To involve the community in *micro-recycling* and regional *macro-recycling*

6.4.1 Objective

The objective of integrated DWM is removal of waste to support humanitarian aid and recycling of materials for reconstruction after disasters. DWM activities are concentrated in the first three phases of the disaster followed by sustainable development (see Figure 6.5).

6.4.2 Emergency phase

The emergency phase starts immediately after the event of a disaster or the end of a conflict. The main objective of DWM during the emergency phase is debris removal and control of waste to facilitate the humanitarian aid, including:

- Organize DWM and planning of immediate waste removal and disposal
- Initial assessment of waste amounts and high priority DWM challenges
- Emergency removal of debris in order to rescue trapped persons and perform demolition of buildings and structures, which exposes people to risks and is an obstacle to humanitarian aid

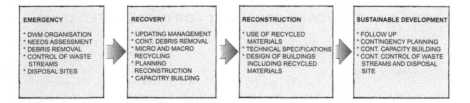

Figure 6.5 Sketch of integrated DWM activities.

Table 6.2 Example of periods of disaster response phases in accordance to the integrated DWM strategy presented in Figure 6.4

Disaster response phase	1st year	2nd year	3rd year	4th year	5th year	>
Emergency						
Recovery						
Reconstruction						
Sustainable development						

- Debris clearance of access roads and clearance of campsites designated for homeless and internally displaced people
- Control of waste stream and mitigation of health risk and danger of infection, including control of disposal of waste materials on assigned sites

In the emergency phase, for instance just after an earthquake, all efforts are concentrated on rescuing people, knowing that persons may survive up to 2 weeks or more trapped in ruins. It is necessary to choose the demolition methods that are most rapid and effective in order to rescue people. On the other hand, serious decisions and actions must be taken very fast to ensure that no further uncontrolled collapse is provoked, thus creating a risk to other trapped people.

Some of the major problems are related to the construction of modern buildings, which collapse in a 'pancake manner', and the appropriate choice of demolition methods. Much heavy equipment is needed, especially lifting capacity. Often, the lack of heavy equipment and lack of demolition experience and knowledge creates time-consuming discussions, for instance the discussion concerning the use of explosives and the risk of resulting ground vibrations affecting unstable buildings, or the initiation of after-quakes due to demolition blasting.

As soon as possible after the disaster, it is recommended that the local authorities should make long-term strategies for the disposal of all kinds of wastes, including demolition wastes from the emergency demolition. If the waste streams are not planned and controlled initially, numerous problems will arise later in the reconstruction phase and much effort will be expended on moving wastes around.

Typically, NGOs will start debris removal using local hauling companies and voluntary resources.

Disposal sites and eventual temporary disposal sites must be planned and assigned. The assignments of disposal sites should be based on contingency plans, or rapid decisions and procedures including assessment of the need for disposal sites, classification and handling of waste in accordance to UNEP/OCHA Guidelines for 'Phase 1: Emergency Phase' (UNEP/OCHA 2011).

It is important that the responsible coordinating aid organizations, typically OCHA, UNDP, International Red Cross and national administrative bodies organize the DWM from the start of the emergency and take responsibility for the following actions:

- Early overview of damages and assessment of the scale of amount of waste and the challenges of debris removal
- Assignment of disposal sites and temporary disposal sites
- Instruments and instructions to NGOs, local administrations, organizations and volunteers on removal and disposal of debris and other kind of solid waste

The experiences from Haiti and Nepal show that the NGOs start debris removal as they see fit. Several months passed before the clusters and working groups responsible for DWM were organized and the first effective steps for debris management were taken.

6.4.3 Recovery phase

After completion of the most important humanitarian aid actions, the main objective of DWM is to continue clearance of damaged buildings and structures and to recycle materials aiming on reconstruction, including:

- Updating the DWM structure and continued planning in accordance with the development of the response scenario
- Continued demolition, demolition clearance and preparing reconstruction of prioritized buildings and structures for reconstruction, e.g. schools, hospitals, roads, etc., including sorting of waste into recyclable materials, non-recyclable materials and hazardous materials
- Introduction of community-based micro-recycling and regional-based macro-recycling (see presentation below), including control of waste streams and disposal of waste, establishment of recycling plants and facilities for crushing debris. Collection and handling of hazardous materials, e.g. asbestos, chemicals, hydrocarbons, etc.
- Planning reconstruction and utilization of recycled materials
- Capacity building of local and regional organizations and private organization with respect to recycling and use of recycled materials

The waste handling should be followed up in accordance to the OCHA/ UNEP Guidelines 'Phase 2: Early Recovery Phase' and 'Phase 3: Recovery Phase'. It is expected that DWM and recycling is organized and controlled. Local contractors mainly perform DWM operations in the recovery phase, eventually supported by international contractors. Based on the early strategic planning of the reconstruction during the emergency phase and recovery phase, the opportunities for use of recycled materials in the reconstruction of buildings and infrastructures should be assessed and discussed. Experiences leaned from Haiti and Nepal show the discontinuities in organization and assignment of key personnel might create problems and delays. Gradually, according to the change of the humanitarian situation, the international staff might be succeeded by other international staff or replaced by local staff. Therefore, it is important to maintain continuity in planning and operation.

6.4.4 Reconstruction phase

Reconstruction of damaged schools, hospitals and other high-prioritized buildings and structures starts as soon as possible, while reconstruction of private houses waits for more detailed planning. Architects and urban planners design standard buildings in accordance to the national requirements and acceptance. Usually, the building design does not consider the opportunities of using recycled materials. New building materials are preferred and reused materials are rejected, because of lack of standards and experience. As mentioned, the UNEP/OCHA Guidelines and the UNDP Guidance Note do not recommend any specific approach for the use of recycled materials in the reconstruction of buildings and infrastructure. Therefore, the opportunities of saving resources and money need a specific strategy for the use of recycled materials in reconstruction of buildings and structures. For instance, the recycling strategy comprising the following activities:

- Assessment of needs for resources and building materials, and opportunities for substituting natural resources and building materials with secondary, recycled resources and materials
- Design of buildings and roads with recycled materials, including evaluation of technical standards for the materials and the testing of the recycled materials
- Instruments and guidelines for reconstruction using recycled materials
- Reconstruction work, including recovering of recyclable materials during demolition of damaged structures and preparation for reconstruction

6.4.5 Sustainable development

In the case of developing countries hit by disasters, the disaster response activities are transformed to sustainable development. The transition takes place over a period of years, depending on available funding and the donors funding policies. In some particularly exposed countries, such as Haiti and the Philippines, it is often seen that a new disaster and response scenario follows an ongoing disaster response scenario. This is the case of Haiti; Haiti had not overcome the effects of the earthquake in 2010 when hurricane IRMA hit the country in 2017.

With respect to DWM, it is important the sustainable development includes:

- Evaluation of the response and follow-up activities
- Contingency planning of waste management systems and facilities, including designation of waste disposal sites and recycling facilities
- Capacity building, including planning and education of local authorities
- Development of waste management and recycling of building materials

6.4.6 Micro-recycling and macro-recycling

Scenarios of single-building transformation and multiple-building transformation are presented in Chapter 5. The scenarios executed in single-recycling projects and multiple-recycling projects are applicable for DWM. However, because of the extent of damage, areal spread and need for response it is recommended that disaster management should be considered in view of the actual priorities, opportunities, resources and funding.

Immediately after the disaster strikes or conflict activities expire, the disaster management will start in the emergency phase on two or more levels:

- Local level, where people will start removing debris by their own means. After some time they will be assisted by national or international NGOs
- Regional or national level, where the UN, NGOs and other international organizations will organize and coordinate the disaster management and debris clearance

After emergency clearance of debris and assessment of the challenges of waste management, detailed planning of waste handling and recycling take place. It is generally expected that organized recycling of debris and other materials can take place late in the emergency phase and at the start of the recovery phase. It is recommended that the recycling activities be organized according to the local, regional or national levels of disaster management concepts presented in Box 6.3.

BOX 6.3   DEFINITIONS OF MICRO-RECYCLING AND MACRO-RECYCLING

Micro-recycling: Manual demolition, sorting, and crushing of debris on-site by the owner/neighbours using mini-crushers or micro-crushers. Reuse of the crushed materials takes place on-site or in the community/neighbourhood.

Macro-recycling: Mechanical demolition, transport of materials to the nearest recycling facility, and crushed by a mobile/semi mobile crusher (capacity up to 250 t/hour). The recycled materials should be sold substituting natural materials at a price compatible with the price of the natural materials.

The opportunities and applicability of the concepts of micro-recycling and macro-recycling have been assessed and discussed in recovery and reconstruction scenarios in Haiti after the 2010 earthquake, and in Nepal after the 2015 earthquakes. See case stories in Section 6.6.

6.5 ENVIRONMENTAL AND ECONOMIC ASPECTS

The economic model, see Figure 6.1 and Table 6.1, shows that the entire resource economy is driven by the marketing of recycled products. The recycled products must fulfil the technical standards and be significantly cheaper than natural material delivered to the construction site. In community-based micro-recycling it is important to identify feasible opportunities for recycling, and it is presupposed that a detailed planning of the use of the recycled materials is prepared in good time before the implementation of the actual debris removal and recycling project.

The discussion of the effectiveness (are we doing useful debris removal?) and the efficiency (are we doing the debris removal in the right way?) requires a closer view of the various conditions and opportunities with respect to access to the buildings to be removed, the community interests with respect to job creation and the further perspectives on recycling and economy. This is a balance between micro-recycling and macro-recycling, effective mechanical operations versus intensive labour work. The choice of technologies must aim at effectiveness as well as considering traditional tools and the future perspectives for introducing new tools and machines. For instance, handheld pneumatic concrete hammers, rather than sledgehammers, should be used for the demolition of concrete structures.

The space for demolition and handling of debris on-site is very critical in many debris removal scenarios. In densely inhabited hilly areas, mechanical equipment and trucks are not feasible. In many cases, the only access to the damaged/destroyed buildings is on foot, and there is often no space for stockpiling close to the demolition site. The debris removal therefore requires many hands, which is rather expensive. The involvement of the private sector is very important, not only from the perspective of sustainable development but also with regard to marketing and reuse of the recycled materials. It is necessary to address recycling technologies that are applicable to the private contractors, and the local building and construction sector in general. For example, the contractors and other stakeholders must accept the reuse of crushed debris as backfill or sub-base in roads instead of river stones. Marketing of recycled materials requires that the recycled materials are cheaper than the materials. For construction purposes, recycled materials must fulfil technical requirements given by the public authorities – for instance, according to American Standards (American Society for Testing and Materials – ASTM). The advantage and competitiveness of recycled debris compared with river stones, limestone, or other materials often depend on the transport costs. The UN and the World Bank's policies regarding disaster assistance emphasis are placed on preventive arrangements and analysis of the 'anatomy' of the disaster and its consequences. It is, as an example, important as soon as possible in the disaster development to aim at a reconstruction of the infrastructure and supply lines of the stricken area.

Disaster waste must be managed according to the principle of Best Available Technology Not Entailing Excessive Costs (BATNEEC). It is very important to remember that emergency actions and short-term activities based on rapid reactions might not comply with long-term considerations and environmental policies. From a purely economic point of view, recycling of building waste is only attractive when the recycled products are competitive with natural resources in terms of cost and quality. Recycled materials will normally be competitive where there is a shortage of both raw materials and suitable deposit sites. Therefore, economical savings in transportation of building waste and raw materials using recycled materials must be considered according to Figure 6.1 and Table 6.1. In a situation of free price formation, the choice between recycled and natural materials depends upon quality. The quality of concrete made with recycled aggregates is the same as, or a little lower than, concrete made with natural aggregates.

In larger post-disaster reconstruction projects, such as Kobe, Haiti and Nepal, as described in the following Section 6.6, the economy will be dominated by transportation costs. These transportation costs involve the removal of demolition products and the supply of new building materials.

The most important opportunities for recycling and disposal of DWM are the following:

- Recycling of debris. Table 6.3 presents options for recycling debris
- Reuse of bricks, steel and wood as a substitute for natural resources in the construction of infrastructure, buildings, bases for shelters, etc.

Table 6.3 List of debris recycling options

Recycling option	Comments
Reuse of stones for reconstruction of buildings	Natural stones of all sorts and sizes. The stones must be sorted before demolition and collected – stockpiling site needed
Costal land reclamation	Only cleaned debris to be deposited. Opportunity for deposit of large quantities of debris. Environmental impact assessment needed
Landscaping, ground levelling Landfilling, improvement of landscape	Depending of the available land and transport distance. The future use of land must be considered. Environmental impact assessment might be needed
General backfill	Many opportunities
Finishing façades, decoration	Using debris fines in mortar and plaster
Base fill for camps	Depending on the shelter planning
Base fill for construction	Planning, design and technical spec. needed
Sub-base for roads and surfaces	Planning, design and technical spec. needed, planning needed
Concrete blocks	Recycling of major amounts requires organization of many production sites, involvement of private sector needed
Aggregates in new concrete	Secondary concrete structures, foundations, tiles, pavement stones, roads, etc.
Gabions, infrastructure	Project and project planning is needed
Backfill support, gabions	Many opportunities, must be planned
Gabions for building structures	'Rubble houses' according to conducted test
Covering of municipal landfills, preparation of new landfills	Synergy between recycling site and municipal disposal site

- Recovery and reuse of valuable building materials, such as windows, doors and roofing materials for reconstruction work, and repair of damaged buildings and construction of temporary shelters
- Reclamation of coastal or rural areas for urban development

The management of debris, especially micro-recycling, implies great employment opportunities. Demolition work, sorting and crushing of materials and transportation of materials requires considerable human resources. This is a very important contribution to social welfare and community recovery.

Most of the recycling opportunities have been internationally tested and accepted. However, not all opportunities are feasible with respect to economic and sustainable debris management. The objective of recycling is to support the overall emergency response and reconstruction of the damaged community. The choice of recycling options shall be based on an assessment of the following issues:

- Effective debris removal (micro-recycling and/or macro-recycling explained above)
- Simple BATNEEC technologies – Implemented by local workers
- Short transportation distances – Overall reduction of transportation costs
- Need for space for working, recycling and temporary stockpiling
- Big quantities to be recycled are preferred
- Private sector interest – Involvement of local contractors
- Community interests – Part of development planning
- Low risk purpose – No risk with respect to future disasters must be introduced
- Certification of recycled materials – Technical standards (ASTM)
- Timing and planning

6.6 CASE STORIES

Box 6.1 presents a list of natural and man-made disasters addressing the challenges of DWM. In the following, observations and experiences of selected earlier disasters are deepened. The case stories comprise the following:

- Great Hanshin-Awaji earthquake on 17 January 1995, Japan
- Clearance and recycling of debris in Beirut after 17 years of civil war in Lebanon, 1995–1997
- Emergency DWM, Mostar after the civil war in Bosnia, 1995–1996
- Clearance and recycling of debris in Kosovo after conflict, 1999–2003
- Clearance of destroyed Nahr el-Bared Camp, Tripoli, 2008–2009
- Strategy for debris management after an earthquake in Haiti, 2010
- Debris management in Nepal after earthquakes, 2015

6.6.1 The Great Hanshin-Awaji Earthquake on 17 January 1995, Kobe, Japan

The great earthquake disaster that struck Japan's Hanshin-Awaji region on 17 January 1995 took the lives of more than 5,400 people and injured approximately another 40,000 people. As the earthquake destroyed the entire social infrastructure of the city of Kobe and its surrounding areas, removal and disposal of earthquake waste became a key issue. The total amount of waste was estimated at 15 million cubic metres (20 million tons), and full-scale reconstruction work was not possible until the waste had been removed. Referring to the proceedings of the Earthquake Waste Symposium in Osaka, 12–13 June 1995 (UNEP 1995) and (Lauritzen 1996/97), the waste management following the great Hanshin-Awaji Earthquake was discussed among Japanese and international experts.[3]

6.6.1.1 General problems

Generally, the problems arising from the waste generated by the disaster were the following (Kobayashi 1995):

- The removal of debris and other wastes that had accumulated on roads because nearby buildings and other structures had collapsed was the most important issue. Temporary storage areas were secured for provisional disposal. Removal of debris from major roads was completed in about a month, though about 3 months were required to remove wastes from smaller roads
- Although some waste treatment facilities were damaged, fortunately no treatment facilities suffered total or even partial destruction. Thus, the main problems were operations hindered by lack of electricity, water, and other supplies, and the temporarily crippled collection and transport services
- An immense quantity of waste was generated by the demolition and dismantling of buildings and structures. Such wastes were a serious problem, because they contained asbestos and other hazardous substances, and the quantity of waste greatly exceeded the ordinary treatment capacity
- Great amounts of waste contained hazardous substances, contaminating the surrounding soils
- The destruction of sewage systems and the increase of population in evacuation centres and other temporary houses led to serious problems with human wastes and septic tank sludge
- Waste from the reconstruction work, repair of damaged buildings and structures and construction of new buildings and structures presented a big problem

6.6.1.2 Waste disposal

Normally, in Japan demolition wastes are separated on-site and disposed of for reuse, tipping or further treatment. However, in this case most wastes were mixed, especially during the initial stages of the emergency rescue operations. Since the capacity of incineration plants in the area was low, wooden wastes were deposited permanently at final disposal sites and

---

[3] Japan Waste Management Consultant Association (JWMCA), The International Solid Waste Association (ISWA) and United Nations Environmental Programme – International Environmental Technology Centre (UNEP-IETC), sponsored the symposium (UNEP-IECT 1995).

temporary storage areas. Temporary incineration plants were established at the Kobe harbour area.

Concrete and debris waste was disposed of

- As filling material for land reclamation
- At existing permanent disposal sites in each municipal administrative area
- At regional permanent disposal sites in the Osaka Bay for major land reclamation projects, see Figure 6.6

At the time of the earthquake, the inland landfill and seaside municipal disposal sites in Kobe City had a remaining capacity of 15 million cubic metres, which were designed for normal disposal of solid waste over the next 10 years. However, because they have been receiving massive quantities of solid waste generated by the earthquake, their capacity was filled within just one year.

Immediately after the earthquake, the central and prefectural governments requested that the regional land reclamation organization, Osaka Bay Centre, receive earthquake-generated, non-combustible wastes and that the Centre should increase its disposal capacity. The existing capacity was satisfactory with respect to debris waste, but there was no spare capacity for hazardous waste, and new waste disposal facilities had to be established. The transport of waste to the regional landfill areas in the ocean was based on barges (see Figure 6.6), but the transfer stations were damaged by the earthquake, and could not be used until they were repaired.

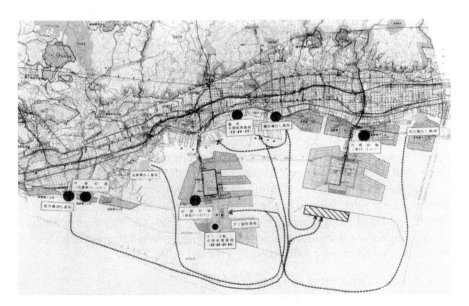

Figure 6.6 Map of Kobe Harbour indicating waste management facilities and routes for waste on barges.

6.6.1.3 Recycling of concrete and masonry debris

Concerning the recycling of debris, it was mentioned that virgin material is cheaper than recycled materials. However, it was also learned that Japan is not rich in natural resources. Therefore, construction materials are rather expensive compared with European prices. Except for recycling of small amounts of waste, opportunities for substituting natural quarry materials with crushed debris in construction work had generally not been utilized. Due to the need for natural aggregates in reconstruction of damaged/demolished structures, it was recommended that options for reuse and recycling of concrete debris should be considered in detail. Reconstruction of port facilities and other parts of the infrastructure needed considerable amounts of natural quarry gravel material of which considerable amounts might be substituted by recycled materials (Lauritzen 1995; UNEP 1995).

6.6.1.4 Summary of important experiences

The state of affairs 5 months after the Great Hanshin Earthquake demonstrated the vulnerability to natural disasters of modern urban densely populated areas, and the capability of the society to recover.

All disasters are unique and they strike the exposed society in different ways. Industrially developed societies are better equipped to establish an effective and rapid emergency response system than developing societies. Earthquakes occurring at night strike the societies in another way than earthquakes occuring during the day when it is light. Therefore, the emergency response must be planned and implemented in each individual case.

6.6.2 Reconstruction of the Central District in Beirut, Lebanon

After 15 years of civil war, the reconstruction of Beirut started in 1994. In June 1994, demolition of buildings in Beirut began in order to implement the reconstruction work of Beirut Central District (BCD). By the beginning of 1996, most of the demolition work was completed, and 10 years later most of BCD was reconstructed (see Figure 6.7).

The clearance of destroyed buildings in BCD was followed by recycling of debris, which were utilized in the reconstruction work. The total amount of debris generated from clearance of BCD from 1994 to 1999 was estimated at approximately 4 million cubic metres. The following opportunities for recycling of debris were examined (Lauritzen 1996/97):

- Sub-base and subgrade material, etc.
- Aggregates for concrete
- Back fill and fill for land reclamation for the Mediterranean
- Scrap metal, to be sold and exported from Lebanon

Figure 6.7 Beirut Central District in 1993 before reconstruction and after reconstruction in 2006.

The options for use of recycled materials required that the reused material should fulfil the requirements and technical specifications for natural materials. Furthermore, it was required that reusable materials should be available in the right quantities at the right time. Therefore, optimal use of the generated material depended on management of the resources. The technical background for the use of recycled aggregates in concrete was based on Danish experiences and European research and development performed by the RILEM Technical Committees on demolition and reuse of concrete and masonry (Hansen 1992; Kasai 1988; Lauritzen 1993).

6.6.2.1 Costs and benefits of recycling

From a commercial point of view, the total cost of the demolition and disposal of debris had to be minimized. Therefore, the feasibility of a recycling initiative was assessed on the basis of consideration of all activities and costs involved in the integrated waste management system, including environmental and other aspects. The debris management system of BCD was based on the business model presented in Figure 6.1 and Table 6.1.

Based on the unit prices used and the estimated amount of debris, as well as the possible demand for reused materials, it was assessed that recycling of materials would save costs, and that recycling would decrease the traffic intensity by 100,000–130,000 truckloads in the first period of the development and reconstruction programme, from 1994 to 1999. SOLIDERE, responsible manager of the reconstruction of BCD, procured a Krupp Hazemag Crushing Plant, which began operations at the end of June 1995 (see the photo in Figure 4.12 in Chapter 4). Initially, the crushing plant did not work properly and economically because of a poor organization and allocation of responsibilities combined with lack of crushing experience. The problems were the following:

- The demolition contract and the recycling contract was divided into two different contracts, without detailed coordination between the contracts
- The demolition contractor was paid a contract based on unit prices per weight debris delivered to the crushing plant. No sorting of waste or selective demolition was required, which lead to deliverance of mixed soil, waste, reinforced concrete and other materials to the crusher plant
- The recycling contractor was not aware of the need for proper sorting of debris and cutting of reinforcement bars before feeding the crusher. Therefore, the crusher was blocked very often and it took an absurd amount of time to clear the feeder. Furthermore, entering the winter season with lot of rain, the wet soil also contributed to many blockages

After learning the lessons and improving the procedures and sorting the debris before loading the crusher, the recycling went on as planned. The recycled materials were used for land reclamation in the sea outside Beirut, creating an area of about 26 ha, which is a new part of Beirut that can be observed today.

6.6.3 Emergency building and solid waste management, Mostar

After more than 4 years of civil war in Bosnia and Hercegovina, the Dayton Agreement of 1995 put an end to the hostilities, which prepared the ground for normal life in the area. Before normal life could be restored, extensive rehabilitation and reconstruction activities were implemented (Lauritzen 1996/97; Yarwood 1999).

In 1995–1997, the EU Administrator of Mostar (EUAM) implemented the clearance of the war damaged city comprising:

* Emergency protection and rehabilitation of buildings
* Demolition work
* Management of materials and waste

6.6.3.1 Emergency protection and rehabilitation

The emergency protection and rehabilitation of buildings comprised all kinds of partial demolition and construction work in order to protect buildings against further destruction, and to protect people against the risk of structural collapse and falling objects. Generally, the protection work comprised all protected buildings (historical buildings) exposed to damage excluding those that had to be demolished due to the extent of the damage.

Some of the most important buildings of historic and cultural interest were the buildings from the time of Austrian-Hungary Empire (before World War I). A *concept for emergency protection and rehabilitation of preservation-worthy buildings* was developed and approved by the representative of UNESCO, Dr. Collin Kaiser.

The concept comprised the following steps:

* Clearance of the building, including removal of garbage and debris
* Careful partial demolition of loose and unsafe structures
* Securing and supporting of the walls and bearing structures to be preserved
* Fencing and sealing the buildings ready for later reconstruction

The protection work was planned and conducted according to the engineering design of the individual buildings with respect to the options for repair and reconstruction. The engineering design was based on a detailed survey of the structural stability, and risk analyses of structural failures and collapses, including the risk of seismic impact. In some cases, the protection work was performed as emergency demolition/protection work comprising the most urgent precautions to secure the building and the public against any hazards. An example of protection of a damaged historic building is shown in Figure 6.8.

Figure 6.8 Historic building in Mostar cleansed and protected for later reconstruction.

6.6.3.2 Demolition work

The demolition work comprised both complete demolition of some structures (removal of foundations and site clearing to prepare for new constructions), and partial demolition of other structures before reconstruction and repair. For the entire city of Mostar, the total demolition work comprised a total gross floor area of approximately 200,000 m². The quantity of demolition waste roughly amounts to 200,000 tonnes. To this amount should be added wastes arising from the reconstruction works of other damaged buildings. The demolition of the damaged 11 storey Razvitak building required careful demolition methods and work because of critical structural explosion damage in the lower floors (see Figure 6.9). The demolition was executed by the use of mini-machines with breakers from the top of the building.

6.6.3.3 Management of materials and waste

The management of materials and waste comprised handling of the following materials categories (see Figure 6.10):

- Reusable materials (roofing tiles, decoration stones for façades, timber,
- doors, windows, construction steel I, U, L and T profiles, iron, etc.   ~ 30,000 tons
- Debris (masonry, stones and concrete), crushed and recycled   ~ 160,000 tons
- Non-recyclable waste materials (organic waste, paper, plastic, garbage)   ~ 10,000 tons
- Hazardous waste including unexploded ordnance (UXOs)

Figure 6.9 View from Hotel Ero in the eastern part of Mostar with the damaged buildings on the confrontation line in the foreground, and the damaged Razvitak high-rise building in the background. The demolition of the building is mentioned in Section 6.6.3.2.

- Soil from excavation of foundations, etc., used for fill. Amounts not included

A sorting and crushing plant was established for recycling debris, mostly for road materials and production of concrete foundation elements, as shown in Figure 6.11.

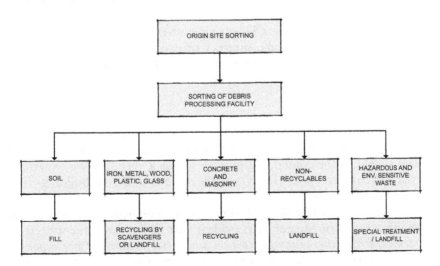

Figure 6.10 DWM processes during reconstruction of Mostar, 1995–1996.

Figure 6.11 Foundation block made by RAC from crushed debris in Mostar in 1996.

6.6.3.4 Implementation of demolition contracts

It was the general policy of the EUAM Dept. of Reconstruction that the demolition work should start as soon as possible, either based on a 'rolling' contract after international tender, or based on several minor contracts to be given to local contractors. Local contractors and local workers should be employed as much as possible. It was the policy of the EUAM that the demolition work should consider the interests of both parts of the city (East and West Mostar) as far as possible. According to meetings with the respective municipalities, the need for demolition was discussed. The municipality of East Mostar had prepared detailed plans with priority given to demolitions and partial demolitions. The EUAM and the municipality of East Mostar had agreed that the physical condition, historical value and ambient situation should form the priority of demolition. The municipality of West Mostar, which did not have the same urgent need for demolition, had not prepared specific plans or expressed a specific priority for demolitions. Finally, after international tendering according to the EU rules, the demolition contracts were executed by Danish and Spanish demolition contractors, sponsored by the Danish Government.

6.6.3.5 Results

The results of the emergency demolition work have been very satisfactory, because the peace process and the firm intention of reconstruction were clearly visible to the contending parties. It was a very positive experience for the citizens and for UNESCO to save more than 20 old buildings from complete destruction.

Based on the experiences of the present emergency protection and demolition work in Mostar, it is evident that there is an urgent need for the *concept for emergency protection and rehabilitation of preservation-worthy buildings* in future emergency and recovery programs after disasters and conflicts, especially the reconstruction of urban areas in Syria.

6.6.4 Building waste management, Kosovo

During the war in the spring of 1999, considerable damage was caused to buildings and structures across Kosovo. More than 120,000 housing units were damaged in Kosovo's 29 municipalities and it was roughly estimated that the waste from damaged buildings and structures at that time reached a magnitude of 10 million tonnes. Based on the experiences of support to the EU Administration of Mostar, 1995–1997, the Danish government issued an action plan in 1999 for Danish support to the Western Balkan, including a programme on Urban Solid Waste Management System (USWMS) and Building Waste Management System (BWMS). The sequence of BWMS projects from August 1999 to December 2003, described in the exit report by the Danish company DEMEX[4] (DEMEX 2004) to the Danish Ministry of Foreign Affairs, comprised a number of activities:

- *Building Waste Management* comprising planning, organization and establishment of a local organization for project implementation, called Waste Demolition Recycling (WDR)
- *Demolition of Buildings and Structures* comprising the demolition of family houses, police stations, barracks and other special structures
- *Recycling of Building Materials* focusing on crushing of concrete and masonry, and the use of recycled materials
- *Assessment of Damaged Buildings*, including damage assessment of major public buildings
- *Hazardous Waste Management* focusing on demolition of sites contaminated by asbestos materials, mines/unexploded ordnance (UXO) and depleted uranium (DU)
- *Business Development* comprising capacity building, business training, transfer of Danish government owned assets to a private Kosovo company, monitoring and follow-up

6.6.4.1 Building waste management

Following the end of the conflict in Kosovo, the action plan focused on three main elements:

- Short-term measures (emergency aid)
- Medium-term measures: Reconstruction and transition assistance (Transition)
- Long-term assistance in order to prevent future conflicts through support to 'democratic infrastructure' (Development)

---

[4] Today a part of NIRAS consulting engineers.

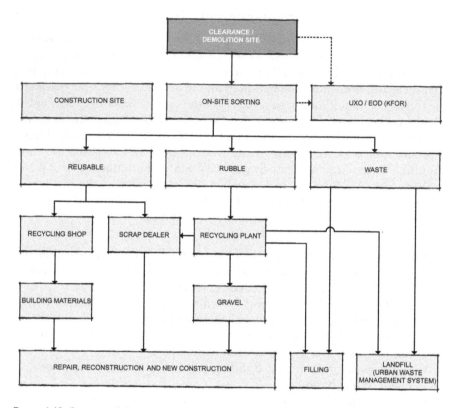

Figure 6.12 Stream of the various fractions of building waste in Kosovo.

The Building Waste Management System (BWMS) included operational and managerial processes and activities related to handling waste materials arising within the building sector, including demolition waste, construction waste and waste from the production of building materials (see Figure 6.12). The demolition, sorting and handling of debris included also dealing with unexploded ordnance (UXO) and explosive ordnance disposal (EOD), which was carried out by the international military forces (KFOR).

The project commenced in January 2000 with the priority of supporting the reconstruction process, especially with regard to the demolition of category 5 buildings (damaged beyond repair).[5] For this purpose, three demolition teams and one recycling team were established. The *demolition teams* were each equipped with an excavator, onto which could be mounted an assortment of hydraulic hammers, concrete cutters and buckets. The teams consisted of up to five workers and one excavator. With the aim of recycling the demolition debris generated from the demolition activities, thereby diverting waste from the landfills, a mobile *recycling plant* was established.

---

[5] According to the classification system of the International Management Group (IMG).

Figure 6.13 Crushing plant for production of recycled materials. Parker Crusher RE 0850 and Viper Turbo 301 Screening plant.

This plant comprised a crusher, a screening unit and a wheel loader (see Figure 6.13). The plant was capable of crushing the concrete and bricks to aggregate material, which was then subsequently sorted into the required size fractions for use in road construction or in low-strength concrete.

Without the opportunities to dispose of this building waste, widespread fly-tipping along roads and in fields is prevalent, causing environmental damage and inhibiting the cleanup work in Kosovo. Therefore, one of the first important steps of implementing the BWMS was to establish suitable depots for recyclable building materials. The aim of the depots was to establish suitable stockpiles of materials collected from the demolition works in the neighbourhood and prepare crushing by mobile crushing plant.

6.6.4.2 Demolition of buildings and structures

Because of the high number of damaged structures and the urgent need for demolition and reconstruction, it was necessary to plan the implementation of the demolition works in details. In order to meet the demands of NGOs, authorities and building owners, special procedures were developed. The procedures focus on approval, planning and demolition, where the approval was the most critical issue.

The issue of property rights has proved problematic for the reconstruction works in Kosovo, since ownership records (cadastre) in numerous cases have either been destroyed or removed from their locations in the municipalities. A central cadastre existed in Pristina and came under the management of a UN habitat project. However, gaining access to the records proved time consuming, since the requests for such access were significant and waiting times of several weeks were the norm.

Since the consent of an owner of a building was needed before the building could be demolished for subsequent reconstruction, a method by which ownership could be verified was implemented within the NGO system. This sought to circumnavigate the potential for buildings being taken over by persons whom were not the true owners, which was a frequent occurrence when one ethnic group had left a village, leaving their homes open for the taking.

The verification system relied on the village leader and at least two neighbours verifying that the claimant was indeed the owner of the house. The claimant also had to document their ownership by showing utility receipts for the same address and name. This, however, could take time and often the NGO would move on to the next house if ownership verifications were not forthcoming.

One further interesting aspect concerns the ownership of the debris generated from the demolition of private homes. The owner was requested to approve the ownership of the damaged building and debris and sign a agreement to WDR before demolition and handling debris. Thus, no owners could later claim compensation for lost income from this sale of debris generated from the demolition of their homes.

The planning and implementation of the demolition work took place based on Western European standards and codes of practice, including health and safety management. Special consideration had to be given with respect to UXO. Therefore, procedures were appointed with KFOR on how to report and handle dangerous items found on demolition sites. Since the damaged houses lie spread throughout Kosovo, often with 5–10 damaged houses in each village, and some of the villages were even shut off for several months during the winter period, then the machinery had to be extremely mobile and robust. Through competitive tendering, therefore, two Fiat-Hitachi 135W wheeled excavators were chosen, each equipped with standard buckets and INDECO 1500 hydraulic hammers. These excavators thus roamed the territory, assisting the house owners with site clearance so that the agencies tasked with rebuilding could start their work. For the demolition of larger structures, such as the Post and Telecommunications (PTK) Building in central Pristina (see Figure 6.14), a Fiat-Hitachi EX 21 was chosen, and the tracked excavator was equipped with an INDECO 2000 hydraulic hammer and Trevibenne F18 concrete pulverizers.

6.6.4.3 Recycling of building materials

Based on previous experiences with the recycling of debris from the demolition of damaged buildings on the former confrontation line in Mostar (Bosnia and Hercegovina from 1995 to 1997), it was decided to establish a similar recycling plant in Kosovo. For this a robust mobile crusher and screening unit were needed, which could travel with the demolition teams and process the debris at recycling depots. After international tendering,

Figure 6.14 PTK building in Pristina, damaged by NATO air bombs. The building was demolished in the spring of 2002. The five-storey building constituted a total floor area of approximately 6,000 m² and generated more than 7,500 m³ (15,000 t) of debris. The work was performed cautiously by a 22 t excavator and one of the 15 t excavators through a combination of mechanical breakage and wire pulling. After reconstruction, the building was handed over to the newly established government of Kosovo in 2003.

the best option for the job was a Parker RE 0850 jaw crusher paired with a Viper 301 Turbo screener (see Figure 6.13). These two machines did short work of any debris generated, operating in often adverse winter and summer conditions and hardly ever breaking down.

During the project in Kosovo, all of the recycled materials produced were sold to both public and private clients. This 100% sales level was only made possible through extensive workshops and seminars demonstrating the use of the recycled gravel for road construction and as engineering fill. The main customers included NATO's forces in Kosovo, who required significant quantities of gravel for their army bases. In addition, local contractors bought the recycled material for construction of roads. Testing the recycled material was not possible in Kosovo since the laboratories had been damaged during the hostilities. Therefore, WDR had to 'export' the gravel for materials testing in neighbouring Macedonia, and then translate the results into Albanian for local use. This troublesome procedure did prove worthwhile in the end, as the recycled materials complied to all the required Yugoslav materials specifications for road construction.

This project has again proved that the recycling of demolition waste generated from reconstruction activities after disasters and wars is a sustainable solution and an important contribution to waste management during these often difficult and chaotic phases. It was estimated that approximately 118,000 t of demolition waste had been generated. A part of the waste coming from remote placed buildings and structures had been used as fill material on-site. Because of long transportation distances, it was not always economically feasible to recycle CDW, or to collect the CDW on municipal depots. It was estimated that approximately 50% of the remaining 90,000 t building waste (i.e. 45,000 t) had been processed at the recycling plant, which generated approximately 17,000 cubic metres of high-quality gravel material.

Generally, recycled materials fulfil European standards for virgin gravel materials. However, in order to compete with the virgin materials, the price of recycled materials must be kept slightly lower compared with the natural materials. In addition, the sale price of recycled products has varied considerably. At the start of phase 1, there was a high demand for gravel materials. On the other hand, the gravel industry was not developed yet, and the most of the gravel material was imported from Macedonia. Therefore, the prices were rather high, and the recycling was very profitable in the early reconstruction phase. Today the prices have stabilized (Figure 6.15).

The sale price of the recycled products is 6–6.5 EURO/m³ compared to the price of 7–8 EURO for natural virgin materials. On average, up to ten persons per day were employed for the recycling function during the operational period of phase 1 and phase 2.

6.6.4.4 Results

The results of the BWMS can be summarized as following:

- Approximately 600 destroyed buildings were demolished and removed making space for reconstruction
- More than 100,000 tonnes of debris and building waste was handled and disposed, of which approximately 50,000 t were diverted from disposal at landfills/dumpsites, and subsequently recycled into gravel material for sale/donation to the reconstruction works

Figure 6.15 UN Mission in Kosovo (UNMIK) depot compound with a ground surface layer of unbounded recycled materials. The surface was of very good quality and was stable, even in very rainy weather.

- Numerous large structures, including two police stations, a bridge and the post and telecommunication building, all damaged from war activities, were demolished and the sites cleared for reconstruction
- The Danish Ministry of Foreign Affairs employed a local work force of approximately 40 workers in the WDR organization and administration from early 2000 until mid-2002. All workers and staff were trained in demolition work, recycling, asbestos handling, scrap handling, business and administration, etc., through their employment, and most of the workers and staff have been employed since the end of WDR in mid-2002
- Capacity building on hazardous waste handling, including asbestos sanitation has been conducted (as well as guidelines for handling asbestos and depleted uranium)

6.6.4.5 Major lessons learned

- It has been proved that in post-war reconstruction there is a need for a BWMS in order to contribute to clearance of war damaged structures and to provide recycled materials for the reconstruction of structures. The BWMS has been found highly suitable for activating local people, since such a system requires a large number of unskilled labourers and thus provides employment opportunities

- The timely removal of damaged buildings and clearing of debris following the end of hostilities provides a positive psychological effect on the society through illustrating that reconstruction works are underway, as well as removing daily reminders of the hostilities. However, the BWMS in Kosovo first became operational in July 2000, a whole year after the cessation of hostilities. Presupposing a quicker response and funding, the BWMS could have been operational at least 6 months earlier, thus providing more assistance to the reconstruction works
- The BWMS in Kosovo has proved effective in providing a link between the relief phase of reconstruction (through solving technical issues) and the development phase of reconstruction (through focusing on economic development). This is a valuable process, which ensures timely transition from humanitarian aid to development assistance
- International organizations, NGOs and other stakeholders showed a very good understanding of and interest in the recycling concept. However, it was problematic to get all, especially the NGOs, to understand that demolition services and recycled materials should be paid for
- Capacity building and establishment of a private recycling business in a post-conflict scenario is a very complex process, and was further delayed due to a general lack of previous experiences, as well as a poor economic and legal structure and culture in post-conflict Kosovo

6.6.5 Clearance of destroyed Nahr el-Bared Camp, Tripoli

In 2007, the Nahr el-Bared Camp (NBC) was the centre of heavy fighting between the Lebanese Armed Forces and the militant Islamist group Fatah al-Islam. A very densely populated area of approximately 200,000 m² (20 ha) was completely destroyed and approximately 30,000 people were made homeless (Figure 6.16).

A major reconstruction project was started in 2008 by the United Nations Relief and Works Agency for Palestine Refugees in the Near East (UNRWA). UNRWA and UNDP entered into an agreement on the management of the rubble removal and explosive ordnance disposal (EOD) project in NBC. Following this agreement, UNDP signed a contract with the local company Al-Jihad for Commerce and Contracting S.A.L. on the provision of services for the demolition, transport and recycling of rubble. UNRWA and the demining NGO, Handicap International, entered into a contract on EOD, which covered the searching and clearance of unexploded ordnance (UXO) and ammunition at NBC. Referring to the UNRWA report (UNRWA 2009) and an article on lessons learned from Lebanon (Lauritzen 2014), the progress of the project is described as follows.

Figure 6.16 Photos of the completely destroyed Nahr el-Bared Camp after fighting between the Lebanese Armed Forces (LAF) and the militant Islamist group Fatah al-Islam in 2007.

The overall objective of the Rubble Removal Project was to prepare the NBC site for the reconstruction of NBC. The rubble removal contract comprised the safe removal and treatment of approximately 500,000 m³ of rubble and waste material in an environmentally sound manner within a period of 18 months. The EOD contract comprised clearance of all explosive items on-site in the rubble above ground surface. The contract was based on an assessment made by the demining NGO, Mine Advisory Group, which showed that the NBC area was heavily contaminated by UXO (see Figure 6.17). Furthermore,

Figure 6.17 Illustration of the challenge of clearance of UXO mixed into the debris.

the Lebanese Armed Forces reported that four to five unexploded air bombs had been launched in the northern part of the area.

6.6.5.1 Challenges

The overall challenge of the Rubble Removal Project was to integrate and optimize as quickly as possible the work processes of the two contractors with respect to the overall goal of the project. The success criteria were to complete the rubble removal and UXO clearance according to the time schedule, 18 months, with respect to the NBC reconstruction project and to avoid any serious accidents.

6.6.5.2 Project implementation

The work started on-site in the middle of October 2008. Shortly after the start it was clear that the rubble removal contractor aimed to complete the project in a much shorter time than the scheduled 18 months. Each of the EOD contractor's four EOD teams included a team leader and four UXO operators. The EOD teams and the rubble-removal teams worked together to remove all rubble, layer by layer, and clear the UXO until the terrain's surface was reached and cleared.

In March 2009, historic remains, comprising stones and marble columns were discovered. According to the findings they represented a large ancient Roman colony and were of archaeological importance. The subsequent rubble removal work at NBC was obliged to consider the protection of the remains and the excavation activities of the Directorate General of Antiquities. The project was completed by the middle of December 2009.

The contractual setup, including the decision to split the rubble removal contract and the EOD contract into two independent contracts, proved crucial during the project implementation. All partners expressed the importance of proper coordination between the rubble removal contractor and the EOD contractor to ensure NBC's successful recovery and reconstruction. However, at the project's inception, the partners did not fully understand the methodology of cooperation and team building essential to working in the field. The rubble removal contractor was very focused on the completion of the rubble removal as quickly as possible. The EOD contractor, with four EOD teams, could not keep up with all the rubble removal teams and provide EOD support, which caused many discussions and frustrations.

The EOD contractors' prioritization of safety in a time-variable contract and the rubble removal contractor's prioritization of work speed due to a fixed-price, time-restricted contract were in disaccord, causing frustration and conflicts of interest throughout the project. The rubble removal contractor allegedly did not understand the requirement of armouring the machines and providing personal protection equipment for demolition workers. Moreover, the EOD contractor often claimed that the rubble removal contractor's personnel did not respect the safety rules. Additionally, due to the safety-distance requirements for rubble removal, allocating work for all four EOD teams on the site was difficult. After positive discussions, the two partners found a suitable *modus operandi* on a daily basis, respecting safety and work performance to successfully complete the project.

By the end of September 2009, UNDP had assessed the total amount of rubble produced by the Rubble Removal Project to be 521,500 m³. 440,000 m³ of rubble was used for land reclamation in Tripoli Harbour (see Figure 6.18), construction of coastal roads and other minor road repairs. Only 6,000 m³ of rubble was destined for crushing, stockpiling and use in the reconstruction of NBC because of the lack of technical quality of the materials. Concerning the recovery of wood, plastic, scrap steel and the

Figure 6.18 Extension of the Harbour of Tripoli with rubble from NBC, 2009.

handling and minimization of non-recyclable waste, the project was very successful.

According to UNDP, 40%–50% of the work force of the local rubble removal contractor comprised local Palestinians. No specific goals on local labour force had ever been set. However, this level of local employment was considered a very successful result of the project. The project, which was estimated to take 18 months from September 2008 to February 2010, was completed before the end of 2009, 16 months after the start of the project.

The density of UXO decontamination was much higher than the assessed amount and density of UXO at the start of the project. A total of seven accidents caused by uncontrolled UXO detonations and two work accidents were registered. One accident was very serious. By September 2009, 11,348 UXO were found and destroyed. Four air bombs (two 250-kg and two 400-kg bombs) were found and handed over to the Lebanese Armed Forces.

6.6.5.3 Lessons learned

It was rather ambitious of UNRWA and UNDP to expect that the Rubble Removal Project should be carried out by two contractors without any specified links between their terms of references (ToR) and contracts, and without a single contracting authority. Splitting the overall rubble removal project into two separate contracts – a fixed-price, rubble-removal contract and a time-variable, EOD contract – was not appropriate. The project setup with respect to the cooperation between the two contractors was problematic, especially regarding safety-measure planning and control, such as maintaining safe distances, wearing personal protection equipment, and so on. In the future, it is recommended that rubble removal contracts and EOD contracts be either merged, with a shared set of contractual conditions, or linked together under full control of one project manager.

Further, rubble removal and EOD are based on different working cultures. Rubble removal, demolition and building waste management are part of the construction sector, while EOD has roots in the military sector and is performed under the terms of the emergency or development sector. The two work routines and cultures should be integrated at all levels. Emphasis should be placed on team building and mutual understanding between the two contactors in order to avoid conflicts of interest regarding speed and safety.

Exploitation of the recycling potential requires a detailed assessment of the inventory of building waste materials and the possible opportunities for the use of recyclable materials. Early planning and detailed requirements of the recycling process in the rubble removal contract are basic issues with respect to successful recycling of valuable rubble materials.

A crushing plant was established for the production of 0–200 mm materials for the Tripoli Harbour landfill. Testing of the crushed materials showed that the materials produced did not meet the technical specification for backfill materials. Therefore, a crushing plant with the necessary sieving facilities must be designed and established so that the materials produced can meet the relevant technical specifications.

6.6.5.4 Overall conclusion

The rubble removal and EOD clearance of NBC was a unique and highly challenging project. All project partners agreed that the project performance had been a learning process. It has been recognized that it is mandatory to have the complete project organization established by the beginning of the project, and that all planning documents, including work plans, health and safety management plans and the quality management plan must be available from day one.

All agreed that the total of seven accidents, including one very serious, was too high. However – in the perspective of the actual high risk – it was fortunate that the number had not been higher.

6.6.5.5 Recommendations

The experiences and the lessons learned from NBC can be used during future reconstruction of war-damaged urban areas, such as will be likely in

Syria following the cessation of civil war (Lauritzen 2014). Daily news from the civil war in Syria during the past years, and reports on heavy fighting using air bombs and all kinds of weapons provide a clear impression of the post-war scenario which is to be expected in many Syrian communities after the fighting ends. The damage to Syria's buildings and the infrastructures contaminated with UXO might look very much like the situation in Beirut in 1990 and in Tripoli in 2008 described in the cases above. It is foreseen that the need for rubble removal and EOD operations will be a challenge to the Syrian people and international aid organizations.

Removal of destroyed buildings contaminated with UXO requires integrated management of rubble removal work and EOD work. Mutual understanding of the work and associated risks, together with open cooperation between the two types of contractors is a mandatory precondition for an effective and successful result (Lauritzen 2014).

Establishing the complete project organization at the project's start is required, and all planning documents including work plans, health-and-safety management plans, as well as the quality-management plan must be available from the beginning.

6.6.6 Strategy for debris management after the 2010 earthquake in Haiti

The 12 January 2010 Haiti earthquake, measuring 7.0 on the Richter scale caused extensive damage to Port-au-Prince and other communities located in the southern part of the metropolitan area. It directly affected 1.5 million persons, and damages were estimated to be $10USD billion. Approximately 600,000 homeless persons sought refuge in temporary camps. Over 250,000 buildings were damaged, approximately 100,000 of which severely. Commercial and governmental properties were damaged beyond repair. The Ministry of Public Works, Transport and Communication (MTPTC) estimated that an amount of approximately 10 million m^3 debris was generated from the earthquake and 90,000 buildings were destroyed (Figure 6.19).

Early in 2010, the UN, US AID, International Red Cross, NGOs, US Army Corps of Engineers (USACE) and many others took the initial steps to manage the most urgent need for removing debris from roads and critical sites.

The earthquake led to serious public health risks to the affected communities, exacerbated by the lack of solid waste collection, leading to waste piles accumulating and polluting urban areas. Furthermore, the waste collection and disposal system were stressed and not able to provide the needed level of service to maintain public sanitation and hygiene. Lack of space for handling, sorting and stockpiling debris was a key problem because nearly all open spaces were occupied with internally displaced persons (IDP) camps.

Figure 6.19 Port-au-Prince after the 12 January 2010 earthquake.

6.6.6.1 *Assessment of the amount of debris following the earthquake*

Two months after the earthquake very few reliable data on debris were available. More than 410,000 buildings were assessed, and the damaged building mass was categorized into five degrees of damage. Based on the damage assessment database, the company Myamoto and UNOPS produced detailed statistics on the houses and the potential of materials, and estimated the total amounts at 9.8 mill m^3 and 10.3 mill m^3. For planning purposes, the 10 mill m^3 figure was applicable for the on-site demolition and debris removal. However, the estimated total quantities of the waste streams to the deposit, handling, and recycling sites were expected to be higher than 10 mill m^3.

6.6.6.2 Review of debris management after the earthquake

Just after the earthquake, substantial initiatives were taken for the urgent removal of debris in order to rescue trapped people and to clear roads for emergency actions. The UN, International Red Cross, NGOs and many others organized debris removal. Because of the emergency situation at that time, the debris removal and the deposit of the removed debris were not coordinated. In February 2010, the US Army Corps of Engineers prepared a proposal for debris management (USACE 2010). USACE and USAID were the responsible partners and members of the Shelter Cluster and Infra Structure & Debris Clusters. In March–April 2010, clearance of debris focused on providence of space for shelters and camps for protecting the homeless people against flood, rain and other climate impact. Also, clearance of debris from riverbeds and channels to prevent flooding of the homeless camps was an urgent activity. During 2010, debris management developed and gained experience. Several debris pilot projects were funded, and the NGOs and private companies organized their debris management processes. In spring 2011, the debris management became mature and effective. All operators, partners and other stakeholders of debris-related activities were organized into the Debris Management Working Group (DMWG) with reference to MTPTC as the overall responsible ministry of the government of Haiti, supported by the IHRC and UNDP.

Much of the demolition and debris clearing work was done manually, because of very poor access to many damaged houses. Local contractors generally rented heavy machinery and trucks. Besides the UK-based NGO Demolition Waste Recovery (DWR) only a few NGO operators have their own mechanical assets. UNOPS organized a major inventory of demolition and debris removal machines including a 250 t/hour crusher, provided by a donor, which was intended to start crushing when a suitable amount of debris had been stockpiled on the site.

The governmental organizations Centre National des Equipment (CNE) and the MTPTC organization Services d´Entretien des Equipments Urbains et Ruraux (SEEUR) had both been working since the day after the earthquake on removing debris, primarily from streets but also from government-owned buildings, offices, schools and hospitals. By the end of July 2011, the amount of debris removed was estimated at 4.0 million m^3. However, no reliable estimates or reporting system on completion of demolishing, clearance and removal of debris from individual buildings and sites existed.

6.6.6.3 Deposit of debris

In February 2010, USACE identified six possible debris-processing sites and four temporary staging areas for debris. At that time it was also known that the Canadian embassy had proposed three landfill sites at the severely damaged urban areas. However, no sites for the deposit of debris was agreed to or decided on. It was more or less up to the individual debris managers to find a suitable site for the uncontrolled dumping of debris.

In March 2010, the World Bank launched a fact-finding mission to Haiti. Its tasks, among others, were to advise the World Bank on the potential strategies for debris management and identification of potential target pilot debris management projects, including a feasibility study on the possible use of the existing Trutier Landfill for handling and recycling. The report of the strategy assessment mission included a proposal for the establishment of a debris management and recycling site located together with the Trutier Landfill (World Bank 2010) (Figure 6.20).

In accordance to the proposal (World Bank 2010), the American company Ceres was tasked to install and operate the Trutier recycling site. The contract started in April 2011 and the site opened for incoming debris by 1 May 2011. The capacity was estimated to 750,000 m³.

The Interim Haiti Recovery Commission (IHRC) assessed the need for one or two more debris management facilities to be placed in the costal area of Port-au-Prince, and has expressed policy on designation of additional dump sites/staging sites (IHRC 2011). In August 2011, UNOPS hired a site on a private area in Port-au-Prince harbour for recycling, and a semi-mobile crushing plant had been installed ready to start crushing in September 2011. In October 2011, UNDP and the Ministry of Environment visited three new possible coastal sites in Port-au-Prince. An old landfill, approximately 3.6 ha with direct access from Boulevard la Saline, was found very suitable for the establishment of a debris management facility.

Important lessons were learned concerning the decision process on designation of deposit sites. It took more than 1.5 year after the disaster to establish the needed number and capacity of deposit and recycling sites because of bureaucracy and the lack of decisions. Private interests and 'not-in-my-backyard' syndromes prevented applicable solutions. Misunderstood perceptions of environmental impact to the maritime and costal environment prevented the establishment of a controlled landfill in the coastal swampy areas and in the Port-au-Prince bay area, such as the solutions found in Japan after the Great Hanshin-Awaji Earthquake in 1995.

6.6.6.4 Gaps and challenges

Compared with other disaster scenarios (see Box 6.1), the challenges and gaps of the 12 January 2010 Haiti earthquake with respect to debris

Figure 6.20 The Trutier Landfill before and after rehabilitation sponsored by the World Bank.

management, as well as the overall response, were many times greater. The overall impact to the communities of Port-au-Prince and nearby regions, the number and extreme density of people, the extreme number of casualties and homeless people, the risk of new disasters and diseases were the overall challenges to the debris management and the implementation of the debris removal activities. The following key issues were identified of importance to the overall debris management strategy:

- *Uncertain amount of debris to be removed*: By August 2011, approximately 6 million m³ remained to be removed according to the official figures.

- *Effective organization and management*: The actual debris management was based on individual project organizations. Organization of the overall debris management system should have been formalized and streamlined.
- *Opportunities for recycling of materials*: There was an urgent need for a clear debris recycling policy.
- *Need for reduction of transport time and cost*: For the debris removal projects in Port-au-Prince, the cost varied from $10–20USD per m^3 depending on the actual distance between the debris removal site and the Trutier Landfill. In some cases, the transport costs were close to 50% of the project cost.
- *Need for recycling facilities*: There was an urgent need for at least one more debris recycling facility near Port-au-Prince downtown with the same capacity as Trutier. A costal reclaimed area for recycling debris was rejected because of environmental objections.
- *Hazardous waste*: It was very important that hazardous waste and waste polluting the environment, e.g. asbestos, PCB, carbon hydrocarbons, heavy metals, chemicals, etc., should have been managed in accordance to international practice.
- *Improved procedures and reporting, including regulations and guidelines*: Procedures for permit to access and demolition/clearance, including procedures for owner acceptance were needed.
- *Monitoring and control*: All debris management activities should have been monitored and controlled in order to ensure the best possible quality, and that the donors and government of Haiti receive value for their money.
- *Information and coordination*: Many organizations, companies, building owners and other stakeholders were involved in debris management and it was a great challenge to keep everyone informed on a suitable level.
- *Funding*: The most important challenge is funding. Gaps in funding followed by gaps in the activities and discontinuities in the overall debris planning and management was a general problem.

6.6.6.5 Recovery and development strategies

The Action Plan for National Recovery and Development of Haiti (PARDN) (Government of Haiti 2010) defined three phases:

- The *emergency* period, which must be used to improve accommodation for the homeless; to return pupils to school and students to university and vocational training centres; to prepare for the next hurricane season in the summer; to pursue efforts to restore a sense of normality to economic life

- The *implementation* period (18 months), for projects to kick-start the future of Haiti and establish a framework of incentives and supervision for private investment on which Haiti's economic growth will be founded
- The *development period* (10 years) during which the reconstruction and recovery of Haiti will become a reality, in order to put the country back on the road to development, followed by another 10 years to make it a real emerging country

6.6.6.6 Debris removal strategies

The IHRC presented the following debris removal strategies (IHRC 2011):

- The Neighbourhood Approach (adopted by the UN, NGOs and American Red Cross) entailing working on debris removal in certain neighbourhoods with the goal of diminishing the various camps populations
- The Sector Grid Approach addresses debris removal by zones, using a 1-kilometre grid and a sub-sector 100 m grid
- Systematic (USACE) Neighbourhood Approach sectors the impacted areas primarily by using department limits, further divided into smaller sectors considering existing political, social, and economic subdivisions

According to the plan, the IHRC recommended a systematic neighbourhood approach. Using the systematic neighbourhood approach, an average cost of $40 per cubic metres removed debris was estimated. To remove 4 million cubic metres by October 2011 (40% of 10 million cubic metres total debris) would therefore cost $160 million. The remaining 60% of debris would have a cost of $240 million, making for a total cost of $400 million, with significant cost-saving potential through strategic goals and management thereof.

6.6.6.7 Sustainable debris management

The Haitian Millennium Development Goal approach (Government of Haiti 2004) Objective 7 *Ensuring a Sustainable Environment* expresses the aim for the elaboration of a strategy for waste management. The strategic plan for long-term development of solid waste management (Government of Haiti 2011) demonstrated the government of Haiti's focus on solid waste management. This includes the justification of a long-term National Debris Management Strategy and the need for coordination between the two strategies.

6.6.6.8 Funding situation

In October 2011, the emergency and the implementation period (18 months), according to the PARDN, had passed. It was expected that the donor interest in supporting Haiti would decrease. In 2010, the debris removal was budgeted with a total of $400 million. Since January 2010, USAID has supported debris management with approximately $70 million. The World Bank funded the establishment of the Trutier Landfill and its operations until the end of October 2011 with $10 million. An extension from November 2011 to May 2012 was accepted by the World Bank. However, the World Bank has other priorities, and does not see opportunities for substantial funding of debris management in the future.

6.6.6.9 Structure of the debris management strategy

The Debris Management Strategy was based on experiences and lessons learnt during the emergency and early recovery period from January 2010 to August 2011. Realistic targets and milestones for debris removal were set according to the principle of Best Available Technology Not Exceeding Excessive Costs (BATNEEC) and the opportunities for funding.

The Debris Management Strategy should comprise two levels: the *Global Level* including planning, management, government relations and donor relations under the responsibility of the national authorities, and the *Operational Level* including the public entities and partners in the field.

6.6.6.10 Recycling of debris

In the greater perspective of the territorial rebuilding and reconstruction of the devastated zones and rebuilding of urban areas, as described in the PARDN, the benefit of recovery and recycling of materials was not expected to be a major contribution to the overall economy of debris management. However, the benefits should be considered in the light of saving resources and reducing the impact to the environment, and the contribution to micro-finance projects. The integrated debris management also benefits the employment of local workers and contributes to opportunities for micro-finance projects. Typical Haitian buildings damaged in the earthquake include private one or two storey houses, multi-storey dwellings, public buildings, commercial buildings, and so on. The main fractions of the debris materials were concrete, masonry, sheet metal and wood (see Table 6.4). In building installations, other types of hazardous substances are found, for instance, refrigerators contain freon. In Haiti, it has not been the tradition to use asbestos roofing materials, which is why asbestos was not expected to be a great issue during the debris management.

In Haiti, the access to natural materials is rather good. Aggregates are picked up in riverbeds, crushed and sorted on process plants inside the Port-au-Prince municipal area. The price of river stones is $21USD/m^3 delivered in Port-au-Prince. This indicates that the production of high-quality aggregates of recyclable concrete debris will not be cheaper than the production of natural aggregates. The dominating factor in the costs of the business

Table 6.4 Distribution of material fractions

Material	%
Concrete, reinforced concrete	50
Masonry	40
Scrap iron, metals, installations, etc.	5
Wood, plastic, glass, paper, cardboard, etc.	5
	100

model is the transport cost. This is true for supply of natural aggregates, crushed recycled materials as well as disposal of debris. The transport cost depends on the time of transport from one site to another. The duration of transport from the downtown Port-au-Prince to the Trutier Landfill took from 20 to 80 minutes depending on time of the day. On average, a truck could only make two trips to landfill areas per day, and the cost of debris transport to Trutier Landfill were $15–20USD per m³. Therefore, it was evident that recycling of materials on-site was the most economically feasible option. Recycling on-site presupposed that all processes (handling, sorting, crushing and temporary stockpiling) are managed on-site or very close to the site. This requires time and space on the site. However, based on field observations it was apparent that locals were recovering steel and other recyclables from the debris removal sites.

Depending on the recycling purpose, the concrete needed to be crushed and sieved according to specific technical specifications. Small mobile crushers (micro- and mini-crushers) and hand-driven crushers were suitable for recycling on-site (see Figure 6.21). For processing of debris on central sites, stationary or semi-mobile crushers were recommended. The production prices of crushed debris from the various crushers are shown in Table 6.5.

6.6.6.11 Recycling options

A long list of recycling options was relevant for the National Debris Management Strategy. Most of the opportunities are internationally tested and accepted (see Table 6.3 and Figure 6.22). However, not all opportunities are feasible for the actual debris removal and sustainable debris management. For instance, high technology techniques such as the reuse of debris in structural concrete were not recommended.

The space for demolition and handling debris on-site was very critical in many debris removal scenarios. In densely inhabited hilly areas, mechanical equipment and trucks are not feasible. In many cases the only access to red tag (unsafe) buildings was by walking, and often no space for stockpiling was found close to the demolition site (see Figure 6.23). Therefore, the debris removal required many hands, which was rather expensive (local

Figure 6.21 Photos of various types of crushers. Mini-crusher (top left), semi-mobile crusher at Trutier Landfill (bottom), and micro-crusher (top right).

Table 6.5 Information on production of crushed materials

Crusher type	Price per unit USD	Production per hour, t	Production price USD per m³
Semi mobile	500,000	250	5–10
Mobile	250,000	60–100	4.5–11
Mini-crusher	43,000	5	–
Micro-crusher	3,800	0.1	30–40

Note: The production prices of the semi-mobile and the mobile crusher include the interval from direct crusher related costs to the total cost including loading, sieving, etc. The production cost of the handheld micro-crusher comprises the workload of three operators, 1–1.5 day per m³, $5–10USD per man per day. The prices and production are approximate and representative.

Figure 6.22 Recycling options. Flagstones for pavements and concrete elements for housing made by RAC (right) and construction of new house on foundation of RAC and reused stones. The timber was imported. The house was built with crossed timber to resist earthquakes.

Figure 6.23 Local debris management.

workers get daily $5–10USD). Rapid removal and recycling of big amounts in big facilities, like the Trutier Landfill, might be preferred depending on the transport distance and costs. Therefore, land reclamation and filling in the Port-au-Prince harbour was proposed as an attractive solution compared to the disposal and recycling at Trutier Landfill.

Since the 12 January 2010 earthquake, the cost of the earthquake removal has been the sum of demolition, clearance, transport and tipping. Since the Trutier Landfill was the only disposal site in the Port-au-Prince area in 2011, the debris removal costs were dominated by transport to this landfill. The income from recycling had not been considered. For optimization of the debris removal, the concepts of micro-recycling and macro-recycling were recommended. The concepts are defined in Box 6.3 and explained by the following examples.

6.6.6.12 Macro-recycling

Referring to the debris situation by the end of 2011, Figure 6.24 presents an imaginary example of macro-recycling in Port-au-Prince downtown and central area, where most buildings and infrastructures were damaged. The example comprised:

- 3 debris removal areas, marked 1, 2 and 3
- 4 landfill sites A (existing site, Truticr Landfill), B (possible new site), C (existing site established by UNOPS) and D (existing site)

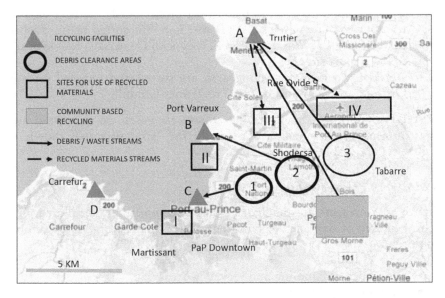

Figure 6.24 Example of the Regional Debris Management system, Haiti –Macro Recycling. (UNDP 2011.)

- 4 sites for possible use of recycled materials I and II (coastal land reclamation and road repair), III (road repair) and IV (airport repair and extension, backfill)
- Area of micro-recycling – to be presented in the next section

Proposed operational strategy:

- Debris for recycling is transported locally to debris collection point. The debris are transported from the collection point to nearest recycling site in order to save money and time
- Wood, scrap metals and other recyclable materials are left in the communities for local recycling
- All organic waste is transported to Trutier Landfill for handling and depositing
- Maximum use of recycled materials in road repair, reconstruction of all kinds of buildings, public and private buildings and land reclamation

Options, for example:

- Debris removal, site 1, is performed mechanically, sorting on-site, transported to recycling site C, sorted again, crushed to max. size 200 × 200 m, transported to coastal landfill site I and dumped
- Debris removal, site 2, is performed mechanically, sorting on-site, transported to recycling site B, sorting again, crushed and sieved according to the US standard ASTM International for road sub-base materials, and delivered to road repair contractor according to specific agreements
- Debris removal, site 3, is performed mechanically, sorting on-site, transported to recycling site A (Trutier Landfill), sorted again, crushed to backfill 0–1 according to ASTM for backfill materials, and delivered to the airport site, IV
- Debris from micro-recycling area are sorted and recycled on-site. Organic wastes are transported to Trutier Landfill

Preferred options and decisions should be based on economic and environmental assessment according to the basic principle shown in Figure 6.1 and Table 6.1.

6.6.6.13 Micro-recycling

Figure 6.25 presents an imaginary example comprising a local area with a high degree of damages of buildings and infrastructure. The example comprises:

- 6 local debris handling and recycling areas
- 1 central recycling site
- 3 collection points of organic waste to be transported to Trutier Landfill

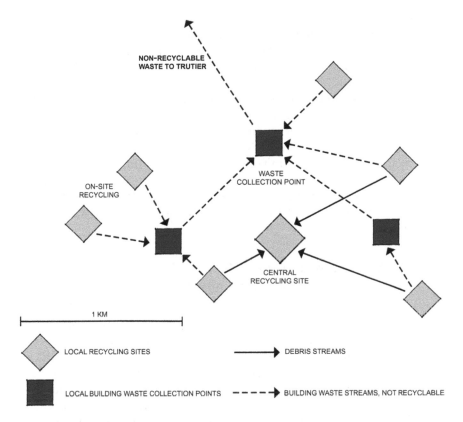

NON-RECYCLABLE
WASTE TO TRUTIER

WASTE
COLLECTION POINT

ON-SITE
RECYCLING

CENTRAL
RECYCLING SITE

1 KM

◇ LOCAL RECYCLING SITES

──────▶ DEBRIS STREAMS

■ LOCAL BUILDING WASTE COLLECTION POINTS

╌╌╌▶ BUILDING WASTE STREAMS, NOT RECYCLABLE

Figure 6.25 Example of Community Debris Management System, Haiti – micro-recycling. (UNDP 2011.)

Proposed operational strategy:

- Debris from private buildings and smaller public buildings are recycled and reused locally
- Wood, scrap metals and other recyclable materials are left on-site and distributed for reuse and recycling in reconstruction work
- Maximum job creation on cash-for-work basis
- All organic waste transported to Trutier Landfill for depositing
- The community will address the opportunities for recycling and reuse of recycled materials

Options, for example:

- Debris removal on six sites, performed manually, sorting on-site, manual crushing of small rubble by micro-crusher or/and mini-crusher, and reuse the materials on-site as backfill or landscaping

- One central recycling site for crushing bigger rubble sizes by mobile crusher, the materials to be reused locally
- Organic and not recyclable waste sorted and collected on three sites and transported to Trutier Landfill for further treatment/disposal

The economic models presents the principles of the entire recycling economy driven by marketing of recycled products for reconstruction work. The recycled product must fulfil the technical standards and must be significantly cheaper than natural materials delivered on the construction site. In community based micro-recycling it is important to identify feasible opportunities for recycling, and it is presupposed that a detailed planning of the use of the recycled materials is prepared in good time before the implementation of the actual debris removal and recycling project.

6.6.7 Debris management in Nepal after earthquakes on 26 April and 12 May 2015

In Nepal, on 25 April 2015 an earthquake with a magnitude of 7.8 on the Richter scale killed over 9,000 people and injured more than 30,000. Its epicentre was east of the district of Lamjung, and its hypocentre was at a depth of approximately 8.2 km (5.1 miles). It was the worst natural disaster to strike Nepal since the 1934 Nepal–Bihar earthquake. A second major earthquake occurred in Nepal again on 12 May 2015 with a magnitude of 7.3, 18 km (11 miles) southeast of Kodari. The epicentre was on the border of Dolakha and Sindhupalchowk, two districts of Nepal. Immediately after the earthquake, different international organizations, NGOs and organizations from the government of Nepal were active in early response and recovery. The safe demolition of destructed buildings, and debris management were among the most challenging tasks.

This case description is based on missions and visits to Nepal in the spring of 2016 by a team of Engineers without Borders Denmark (EWB DK) and Emergency Architecture and Human Rights (EAHR) in order to support the reconstruction of Nepal, in an attempt to implement recycling and the use of recycled materials in reconstruction of buildings and infrastructures (EWB DK and EAHR 2016).

According to the Solid Waste Management Technical Support Centre's Desk Study (UNEP IETC 2015), approximately 14 million tons building of waste material (debris) was generated from the earthquakes in the 14 districts. In a later report made by United Nations Environmental Programme (UNEP 2015) (Government of Nepal 2015), the estimated amount of debris was reduced to approximately 10 million tons. The UN organizations, other international organizations and NGOs worked closely with the local governmental organizations like the Department of Urban Development and

Building Construction in urban areas, the municipality in semi-urban areas and the District Technical Office and Village Development Committee in rural areas. Allocation of the demolition and debris management work was made at the district level.

UNDP played an important role in the demolition and debris management right after the earthquake. Within 3 months, UNDP employed over 3,000 people (43% women) through its cash-for-work programme, and has deployed field teams that include 90 Nepali civil engineers working as UN volunteers, who have been trained by international demolition and debris experts. These teams demolished over 3,000 damaged structures in some of the hardest-hit and remotest areas. More than 170,000 cubic metres of debris have been cleared. International Organization for Migration (IOM) was also one of the key players in demolition and debris management, supporting the demolition of private and public buildings in Kathmandu, as well as in rural areas.

By the end of 2015, the earthquake disaster response transited from recovery to reconstruction. The former cluster organization was changed and the National Reconstruction Agency (NRA) was established. The most urgent debris removal, especially in Kathmandu and other urban areas, had been completed and the focus shifted from debris management to reconstruction. In the short term, bridging gaps between DWM and reconstruction of buildings and infrastructures was necessary for substitution of raw/virgin materials with recycled materials for reconstruction work. The long-term sustainable development and optimization of the resources will characterize the future processes of reconstruction, mitigation, prevention and preparedness.

Most of the damaged concrete buildings in urban areas in Kathmandu and other cities were cleared within the year after the earthquakes, while most of the rural damaged buildings still needed to be demolished/cleared (Figure 6.26). Local workers were employed on the basis of work-for-cash for demolition work in accordance to National Guidelines and UN demolition guidelines. However, it was observed that the demolition works did not always follow the National Guidelines (Government of Nepal 2015) for safe demolition and debris management (Figure 6.27). USAID has contracted a New Zeeland demolition company to train Nepali demolition experts in demolishing high-rise buildings taller than three floors (Figure 6.26).

In December 2015, the Solid Waste Management Technology Support Centre (SWMTSC), UNEP and the Leadership for Environment and Development, Nepal, published the report DWM – *Policy, Strategy and Action Plan* (SWMTSC, UNEP and LEAD 2015). The report presents a set of visions, objectives, strategies and action plans very much in line with the proposed views and strategies on integrated DWM presented in the start of this chapter (see Box 6.4).

Figure 6.26 Damage in rural areas, district Dolakha, damage in village (left), and damaged school building (right).

6.6.7.1 Debris management

Bricks, wooden frames and stones were reused and steels bars were sold as scrap for recycling. Wood not fit for reusing was used as firewood. However, debris management focused more on debris removal than reuse, recycling and recovery (3Rs) of debris.

Handling of debris in many areas was done by dumping the debris in specific sites or over slopes used for landfills, land reclamation and repairing

BOX 6.4 POLICIES, ACCORDING TO THE REPORT OF SWTSC, UNEP, LEAD REPORT (UNEP 2015), PAGES 21–25

PROBLEMS AND CHALLENGES

Large quantities of mixed waste produced within a very short time frame, lack of knowledge of practical technology, lack of coordination, limited funding to handle post-disaster situations and capacity building.

VISION

To achieve a cleaner, greater and safer nation by protecting and managing the environment by resource recovery and safe final disposal, in line with the UN Sustainable Development Goals (SDG) targets.

MISSION/GOAL

With wide-range participation of concerned stakeholders, establishment of a sustainable DWM system including reduce, reuse, recycle, treatment and safe disposal of disaster waste and strengthened coordination among government agencies, the private sector, civil society and other stakeholders.

OBJECTIVES

- Establish an integrated DWM system into a holistic waste management system maintaining health and an ecological balance
- Strengthen and establish appropriate institutional setup with coordination mechanism between all concern stakeholders including National Reconstruction Authority for DWM
- Promote resource recovery of disaster waste using innovative and sustainable technology through public-private partnership models
- Develop infrastructures for the final disposal of risky waste.

POLICY RECOMMENDATIONS

The collective input from the field visit, discussion meetings and workshops, waste management measures, waste minimization, and safe disposal of disaster waste, are as follows:

- Prevention and preparation
- Empowerment/community mobilization
- Response and removal
- 3R principles: Reuse, recovery and recycling
- Monitoring and evaluation

dirt roads and tracks. Debris was used directly without processing and downsizing it further, and was therefore not used in the most beneficial ways. For instance, road repair in Kathmandu was performed with unsorted materials and big blocks, which hampered the traffic. During the visits to three of UNDPs demolition sites and schools in Dhulikhel, the concrete debris were randomly disposed of and no reusing of concrete debris was seen. Operators and transporters of debris removal preferred the shortest possible transport distances.

A draft Debris Management Strategy of Kathmandu Valley had been prepared in 2014, with designation of 83 selected sites for debris management. On meetings with the UN and Nepali Government organizations it was learned, that the strategy had not been effective, and the planned processes for recycling were not implemented (Figures 6.27–6.29).

Figure 6.27 Most demolition in rural areas was performed manually – often neglecting the National Debris Safety Guide.

Figure 6.28 Manual crushing of concrete for recovering of steel bars (top) and cleaning bricks for reuse (bottom).

6.6.7.2 Reuse, recycling and recovery (3Rs)

The international guidelines, presented in Section 6.3, such as the DWM Guidelines (UNEP/OCHA 2011) and the UNDP Guidance Note Debris Management (UNDP 2013), address and recommend reuse, recycling and recovery (3Rs) as an integrated part of DWM. Referring to the Kathmandu Valley Post-Earthquake Debris Management Strategic Plan (Government of Nepal and IOM 2014), 3R is pre-planned. During the first phases of the

Figure 6.29 Reconstruction of a school using recycled bricks.

earthquake response in Nepal, in 2015, usable housing materials, bricks and scrap materials have been collected. However, no organized recycling and reuse had taken place.

A majority of buildings in villages and rural areas are built of brick, wood, stones, and have stone or corrugated iron sheet roofs. There was a considerable potential for recycling concrete, wood, bricks and steel. However, no or very little recycling of debris was seen in the spring of 2016, one year after the earthquake.

Concrete buildings are mostly seen in larger villages, cities and urban areas and are up to 7–8 storeys in height. The EWB and EAHR team observed a number of severely damaged buildings in Kathmandu, Dhulikhel and Charikot which are still standing, indicating an ongoing need for demolition work to be carried out, including management of debris. Many public buildings, hospitals, health posts and schools, such as the schools in Dhulikhel – demolished by UNDP – have high potential of reusable and recyclable materials: bricks, concrete, iron bars and corrugated iron sheets.

The use of crushers for crushing and recycling of debris was discussed with IOM. After the two earthquakes in Nepal, a survey carried out by IOM on the availability of crushers for debris treatment showed that no crushers were available in Nepal. However, on the field trip to Charikot the team visited a crushing plant at the riverside for producing aggregates with a three step crushing process of stones from the river. Concrete and masonry debris has been used for road repair, for example in Kathmandu. However, the unsorted materials tipped in holes in roads without proper layering and compacting is not an effective way to make road repairs. Proper repair using crushed recycled material is needed.

On-site recycling of stones, concrete, wood and steel should be highly prioritized because of logistic challenges. Because of poor access roads and lack of roads within the local communities, the transport of building materials takes a long time. For instance, the driving time from Kathmandu to Dhulikhel took at least 6 hours, and the transport from public road to some of the school visited was only possible on foot. The building materials such as brick, stones, wood, cement and so were transported on foot in 30–50 kg bags over several hours.

6.6.7.3 Reconstruction

In spring 2016, the Nepal government was already carrying out road repair and landslide protection using gabions, filled with materials transported from the riverbeds. Using gabions filled with recycled crushed materials substituting for river gravel is an obvious solution. The following opportunities for recycling and use of recycled materials in reconstruction work were recommended:

- Improved road repair
- Debris in gabions for landslide protection and roadside protection
- Stabilization of reclaimed land
- Crushed concrete debris used as gravel in improved bricks
- Crushed material for rural road and trail repair
- Crushed material in new concrete structures substituting for virgin/ raw materials from riverbeds
- Crushed material and debris in new building (educational building, community centres, governmental building, etc.)

6.6.7.4 Strategy

Referring to the strategic approach presented in the DWM Report (SWMTSC, UNEP, LEAD 2015) and policies presented in Box 6.4, the following strategy for sustainable debris and resource management was proposed:

- Assessment of the potential of resources from debris applicable for repair and reconstruction of damaged buildings and infrastructure
- Establishment of local resource management systems at village and district levels, coordinated with the planned resource centres
- Capacity building, training and education on debris and resource management, at national, district and village level
- Introduction, training and education on the use of recycled materials in reconstruction works
- Introduction of specific technical and administrative requirements and guidelines for 3Rs of debris and use of 3R resources in reconstruction works

6.6.7.5 Operations

The implementation will adopt an integrated and participatory approach (social mobilization) that will contribute to the reactivation of local economies (job creation), particularly through debris recycling activities and reconstruction with recycled resources and local and traditional materials. Local workers are hired on a cash-for-work basis. Stockpiling and recycling sites are installed in the neighbourhoods. Recycling of debris in the neighbourhoods will produce materials for reconstruction and mitigation measures, for instance landslide protection.

Given the topography and access condition of destroyed neighbourhoods, the project will use different equipment (crushers, debris sorters, etc.) installed near the buildings. The size and capacity of these devices will be adapted to the nature of sites to be cleared while combining high-intensity labour from the community with small machinery, as needed. The project will evaluate the costs/benefits of rental or purchase of such equipment.

Recyclable debris will be handled on-site by small and micro-enterprises or NGOs. This methodology will allow the community to be a driving force in the reconstruction of their neighbourhoods, and at the same time it will inject cash directly in the community thanks to the employment of local workers for the recycling of debris and reconstruction. In this sense, rapid assessments of new income generating opportunities will be developed together with small vocational training programs for the beneficiaries, including both men and women. Additionally, support to micro- and small enterprises, not necessarily debris related, could be considered in support to the reactivation of local economies.

The third income generation opportunity will be linked to the possibility of recycled debris. As permanent housing and house repairs initiatives gradually increase, along with the reconstruction of affected neighbourhoods, the demand for processed debris for backfilling, paving blocks and other non-structural purposes will also grow. Agreements with the municipalities and national institutions such as the Department of Road, Department of Education or Department of Urban Development will be pursued for establishing solid partnerships. National institutions could use crushed debris for infrastructure works such as roads and canals. This will most likely lead to an additional source of income, thus generating activities for project beneficiaries.

6.6.7.6 Economy

The overall requirements of the DWM, 3R and reconstruction projects are:

- They are economically feasible
- They do not expose the owners of the structures, third parties or the environment to any unforeseen risks

The budget of DWM and using recycled materials in a specific reconstruction project must be calculated in accordance to the actual preconditions, including the actual cost of transportation and need for resources.

The EWB EAHR team proposed business models such as those presented in Section 6.1 (see Figure 6.1 and Table 6.2) and recycling scenarios like macro-recycling and micro-recycling, which were also proposed in Haiti (see the Haiti case above). The economic model was based on the use of micro-crushers and mini-crushers used in Haiti from 2010, as shown in Figure 6.21. Besides crushing debris from damaged buildings and structures, the micro-crusher is very useful for crushing natural stones and rocks into sorted gravel for general use in local communities. However, there is no evidence that the recommendations have been followed.

Chapter 7

Construction, demolition and DWM in the future

Should the global population reach 9.6 billion by 2050, the equivalent of almost three planets could be required to provide the natural resources needed to sustain current lifestyles.

UN SDG Goals 2030, Goal no. 12

7.1 INTRODUCTION

Compared to other sectors the construction sector is conservative. No substantial innovation in the construction sector has taken place for the last 50 years. Most of our buildings and constructions are designed according to more than 100 years of tradition, and concrete and masonry are the dominating materials. According to the World Economic Forum (WEF), (WEF 2016), the construction industry accounts for about 6% of global GDP and is growing. It is the largest consumer of raw materials and other resources, using about 50% of global steel production and more than 3 billion t of raw materials. Constructed objects account for 25%–40% of the world's total carbon emissions. Any improvement will have a major impact on sustainable construction, not least on recycling of CDW.

This chapter deals with future construction resources, looking at the future challenges, megatrends and opportunities focusing on targets relevant to integrated construction, demolition and DWM.

7.1.1 Challenges

Looking at the future of the construction sector, the following are some of the most important challenges:

- Growing population and the need for housing
- Aging buildings and infrastructure
- Need for environmental, social and economic sustainable constructions
- Growing concerns of natural disasters

The world's urban population is expected to exceed 9 billion by 2050. It is expected that the continued migration from rural to urban areas and the concentration of people in megacities accelerate the need for housing. The need for housing in urban areas – where the construction process is very complex, owing to space constraints – is accompanied by the need for increased and changed infrastructure facilitating traffic, water supply, sanitation, and so on.

Aging buildings need renovation or transformation into new buildings. A great part of the buildings constructed after World War II in Europe, especially in the 1960s and 1970s were build according to requirements of urgent need and speed. The development of the oil countries in the Middle East required rapid urban development and housing. For instance, by 1990 most of the building stock in Kuwait was younger than 20 years.[1] Looking at the urban development around the world, especially in the Far East 'tiger' countries[2] and BRIC countries,[3] one notices the accelerating need for housing and the required construction activities targeting quantity before quality. Generally, there is huge need the world over for renovation and transformation of the existing building stock. The change of industry from heavy labour-intensive industries powered by coal and oil to light, digitalized and less resource demanding industries powered by natural gas and non-fossil energy requires transformation of the huge old stock of outdated and obsolete buildings and structures. Ageing infrastructure assets in developed countries demand proper maintenance, upgrading, replacing or newly built assets; there is, of course a fast-growing societal need for infrastructure assets in emerging markets. So overall, there is immense opportunity, and responsibility, for the construction industry. According to the WEF, one other notable development is the increasing number of infrastructure megaprojects.

Natural disasters and conflicts continue to challenge exposed regions around the world. The expected change of climate and the rise in temperature of the oceans increase the risk of higher frequency of natural disasters with more severe impacts to the affected societies. Therefore, the construction industry also has to address the growing concerns over natural hazards (notably, flooding, hurricanes and earthquakes), and to enhance resilience. So, new emphasis is being placed on devising risk-mitigating solutions, especially in urban areas with high population density.

Increasingly, sustainability is becoming a requirement rather than just a desirable characteristic, and its pursuit is bound to affect both the construction process and the built asset itself. The construction sector produces an enormous amount of waste, so the more efficient use and recycling of raw materials, even a small improvement, offers huge potential benefits. Other new priorities are emerging accordingly, including optimizing

---

[1] During assessment of building waste in Kuwait in1990 performed by the Danish consulting company Ramboll – just before the Iraqi invasion – it was mentioned that buildings of more than 20 years were very old.

[2] Singapore, South Korea, Hong Kong and Taiwan.

[3] Brasilia, Russia, India and China.

space, for example, and ensuring more efficient methods of heating, cooling and lighting (WEF 2016).

7.1.2 Megatrends

Referring to the WEF, the construction industry is concerned with the health and safety not only of workers but also of the people who actually live or work in the buildings. Employee health and productivity are linked to the quality of the indoor environment, and that quality is largely determined by decisions made during project development and construction. The construction sector's responsibility does not end with the delivery of the project: the entire operations or use phase is affected by the initial selection of materials. The safer the materials, the better it is for health and the environment. For instance, asbestos has been outlawed in many countries as a construction material, and construction companies are increasingly motivated to ensure that the living and work environments they create are ergonomic and allergy-free. In addition, at the end of a building's life, safer materials can be more easily integrated in the circular economy. Taking inspiration from the WEF (2016), the Ellen MacArthur Foundation (EMF 2014), Ecorys (2014), EU commission (EU 2014) and the Danish report *Building a Future* (Jensen and Sommer 2016) the following trends, among others, are mentioned:

- Advanced building and finishing materials
- Applying industrial production processes
- Expanding the reuse and high-quality recycling of building components and materials by applying design for disassembly techniques, material passports and innovative business models
- New construction technologies, e.g. 3D printing
- Standardization, modularization and prefabrication
- Smart and life cycle–optimization equipment
- (Semi-)automated construction equipment
- Increasing the utility of existing assets
- Extended lifetime of constructions
- Improved processes by the use of Building Information Modelling (BIM), big data, virtualization, etc.

7.2 FUTURE TRANSFORMATION AND RESOURCE MANAGEMENT

Looking at the model of transformation of life cycles, presented in Figure 7.1 (according to Figure 2.2), our major challenge for the future integrated management of CDW is transformation of the old existing stock of structures, and extension of the existing stock with new structures. The

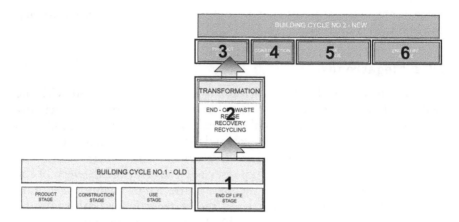

Figure 7.1 Coupled building cycles enabling transformation of waste from building no. 1 (old) to resources for building no. 2 (new).

stock of structures comprises dwellings, industrial buildings, infrastructures and other structures related to the building and construction industry.

The trends and future aspects until 2050 are described as follows, referring to Figure 7.1:

1. End-of-life of existing old structures, including demolition and CDW handling
2. Transformation of old structures into new structures, including EOW and transformation of CDW to resources
3. Design of new structures
4. Construction of new structures
5. Use of new structures
6. End-of-life and demolition of future old structures

7.2.1 End-of-life of existing old structures

The general trends are that the existing buildings of good quality will be rehabilitated and lifetime-extended, while the outdated and poorly maintained building stock will be demolished and replaced with new buildings. Today, the general perception of lifetime of buildings in Europe is 50 years depending on their quality and purpose. The life cycle of housing buildings depends on the quality and use of the construction combined with the cultural and social needs together with safety and environmental demands. In Europe, a great number of old buildings from the eighteenth century and the early nineteenth century are still in use and are expected to remain so for many years in the future. The reason for this considerable lifetime length is the sound and durable materials (bricks, wood, iron and concrete) and simple construction methods. In contrast, many buildings from the 1960s and 1970s were constructed with industrial prefabricated concrete elements. The elements were assembled and coated with various compounds containing substances hazardous to the environment, typically asbestos, lead and other metal-compounds, hydrocarbons, poly-chlorinated biphenyls (PCB) and polyvinylchloride (PVC) and others (see Chapter 2, Section 2.6). It turned out that urban development of this period was not in harmony with the cultural and social development. Furthermore, the indoor climate and fire safety of the buildings did not fulfil the updated requirements. Therefore, a great part of the building stock from this period needs to be transformed by demolition or renovation to new, greater and better dwellings after a short lifetime of less than 50 years. Many extensive urban developments have taken place all over Western Europe, as mentioned in Chapter 2 and 3, such as Ballymun in Dublin, Ireland. In Denmark, the government has set up a plan for the rehabilitation of a number of ghetto areas. It is expected that the transformation of the poor buildings from this period will continue for many years in the future.

Because of the movement of the population to urban areas many settlements, houses and urban areas are left empty (see Box 7.1). The empty buildings are a huge problem in all industrial countries, which calls for public action and demolition. In contrast to the problems of the empty buildings, we face the challenges of maximum utilization of the extremely expensive areas in downtowns of megacities. High-rise buildings are demolished for the construction of higher buildings. For instance, JPMorgan Chase announced in February 2018 that it will demolish its 52-storey headquarters on Park Avenue in New York and build a new 70-storey headquarters on the site for 15,000 employees.

Following the industrial development over the past 50 years, a great part of the old heavy industry buildings have been demolished or transformed for new purposes. Most of the heavy industry has moved from the expensive and crowded city centres to new cheaper areas in the outskirts or developing regions with access to cheaper working power (see example of Carlsberg, Denmark, Chapter 5, Section 5.5). This trend will continue, and the frequency of moving the industry away from the cities will increase.

The rising need for infrastructure calls for transformation of structures. The transformation systems requires exchange and extension of roads and bridges, such as replacement of the old Storstroems Bridge in Denmark and the Oakland Bay Bridge in California with new bridges, as mentioned in Chapter 3. The maritime traffic has concentrated in big ships and terminals and much of the maritime traffic has transformed to land- and air-based traffic. This made many harbours superfluous, and many local city harbours have moved to new places. In the future, before we reach 2050, the transportation systems will change. It is also foreseen that automatic trains and cars need special lanes and facilities, which will require considerable changes of infrastructure. In addition, the energy sector has transformed and these transformations will accelerate because of the need for reduction

BOX 7.1 CHALLENGES AND MEGATRENDS: FACTS AND FIGURES

- By 2050, the world population will exceed 9 million
- Half of humanity – 3.5 billion people – live in cities today
- By 2030, almost 60% of the world's population will live in urban areas
- 65% of the next decade's growth in construction will happen in emerging countries (WEF)
- No 1 consumer of global raw materials is the construction industry (WEF)
- 90% of the 2050 built environment already exists (EU)
- 50% of the solid waste in the United States is produced by the construction industry (WEF)
- Three times as many disasters were reported last year as in 1980 (WEF)
- 95% of urban expansion in the next decades will take place in developing world
- 828 million people live in slums today and the number keeps rising
- The world's cities occupy just 3% of the Earth's land, but account for 60%–80% of energy consumption and 75% of carbon emissions
- Rapid urbanization is exerting pressure on fresh water supplies, sewage, the living environment, and public health but the high density of cities can bring efficiency gains and technological innovation while reducing resource and energy consumption
- The EU27 consumed between 1,200–1,800 million tonnes of construction materials per year for new buildings and refurbishment between 2003 and 2011 (Ecorys)
- The biggest fraction is aggregate materials, which represent about 45% of the total materials by weight. Concrete, with 42% is the second next fraction by weight, followed by bricks with 6.7%. The largest metallic fraction is steel, which accounts for about 2.5% of the materials use by weight. Wood (timber) which is the largest biotic fraction, accounts for around 1.6% of material use. The rest of the materials (including copper, glass, aluminium, etc.) each make up less than 1% of material use (Ecorys)
- Transport cost in the extraction sector accounts for around 13% of the total costs, which makes it uneconomical to transport the materials further than around 35–50 kilometres (depending on the diesel prices) (Ecorys)

Source: UN Sustainable Goals 2030 (SDG), goal no. 11 (UN 2015), World Economic Forum (WEF 2016) and (ECORYS AND CDI 2014).

of greenhouse gases. Less resource demanding and non-fossil-based power production will substitute for the old fossil fuel-based power plants. During the past decade, we have seen the offshore oil industry changing to maritime wind turbine farms because of the lack of oil and gas in known exploration fields. The dismantling of the old offshore installations is beginning with removal of the Ekkofisk and Frigg fields in the North Sea and a number of fields in the Gulf of Mexico. Huge challenges of dismantling and removal of the remaining installations remain ahead. By the end of 2017, Mærsk Oil and partners have decided to carry out a complete renewal of the Thyra Field in the Danish sector of the North Sea, with a budget of approximately 3 billion US$. In general, the lifetime of wind turbines is expected to be 20 years. This indicates an impending need for decommissioning of the early generations of wind turbines and handling the considerable amounts of fibreglass from the wings and concrete from the foundations (Gjødvad and Ibsen 2016). Decommissioning of old wind turbines has started. The decommissioning of the old stock of nuclear power plants is ongoing. However, the unsolved problems of sustainable handling and disposal of radioactive waste are challenging. Short-term disposal solutions are accepted.

7.2.1.1 Demolition in the future

Today, specialists perform demolition and CDW management according to independent contracts without strong links to planning and construction of new structures. In accordance to the subject of this book and the EU initiatives discussed in the following section, it is expected that demolition, CDW management, recycling and transformation of buildings and materials will be integrated in the building and construction industry successively during the coming decades.

Referring to the trends of development of demolition technologies and methods during the past decades, the following trends are observed:

- Demolition work will be less manual-work demanding. In order to reduce the risk of workers' health and safety, the industrial work will be maximized and the manual work at the demolition sites on structures will be minimized
- Robot technologies, remote, automatic (semi-automatic) controlled machines and equipment will be used to promote the speed and efficiency of the demolition work
- Special technologies for demolition of high-rise buildings will be developed, such as long-reach machines and top-down demolition systems. Requirements for demolition of high-rise structures by explosives will be more rigorous
- Special technologies and methodologies for protection of demolition sites will be developed to meet rising public requirements for control of dust, noise and vibration with respect to the local impact of effects

from demolition, and to protect the health and safety of the demolition inside the workplace

- Further development of equipment for environmental cleansing and sorting materials will take place
- Selective demolition on-site will be accompanied by dismantling of building elements, which are removed, downsized and sorted mechanically on local sorting points. It is estimated that industrial treatment of some type of complex elements, e.g. sandwich elements, are preferred instead of on-site demolition

7.2.2 Transformation of old structures into new structures

Some decades ago, the trends of urban development were based on demolition of all old buildings without special regard to the quality and type of old structures. Today, we are much more concerned about old buildings of cultural significance and high quality and try to adapt the old structures into the sustainable design of urban development. Apart from the poor- and low-quality structures, which have reached the end-of-life and demolition, the building stock will be maintained and renovated. Dwellings are renovated to meet requirements today with respect to size of area, energy consumption, facilities, and so on. Some buildings are renovated for other purposes, such as old industrial buildings, as described in Chapter 2, Section 2.5. Following the demand of housing and offices in city centres, many cities have expanded the limits of height of structures and allowed height-extensions of existing buildings. Today, we also see renovation and extension of relatively new high-rise structures. Figure 7.2 shows the planned renovation and extension of a high-rise building in Sidney. According to information from 3xN Architects, the renovation comprises extension of the original floor area from 1100 m^2–2200 m^2, retaining 65% of existing structural columns, beams and slabs, and 98% of structural walls will be retained. It is expected that the work will be completed in 2021. The owner is AMP Capital and partners are Arup and BVN. The goal of sustainability in design is certified according to BREAM and LEED. Façade and slabs will be designed for disassembly (3XN 2018).

It is expected that renovation and extension of existing of buildings will be an important trend of in the future transformation of structures.

7.2.2.1 Transformation of CDW into resources

Beyond doubt, the trends of sustainable and socially responsible construction today will be implemented in the future transformation of existing structures. All degrees of transformation of structures, from total demolition to renovation and overall rebuilding will produce waste materials.

Figure 7.2 Renovation of high-rise building in Sidney. Top left: Shows the old building marked with a dark frame. Top right: Shows the design of the new building. The models show the building before renovation, and after partial demolition and completion of the new building. (Photos and models, 3xN Architects, Copenhagen.)

Building owners and developers will understand the challenges of reduction and recycling of both demolition and construction waste materials. They will take necessary actions to exploit the opportunities and overcome the barriers following the recommendations presented in this book.

It is expected that:
- Sustainable and socially responsible design and construction will be implemented in future transformations of existing structures, including prevention, reduction and recycling of CDW and resource efficient integrated resource management
- Controlled processes are invented and introduced ensuring traceability and documentation of the streams of CDW and transformation of CDW into valuable resources

- Effective and accurate sorting systems are developed based on high technology sensing and separation systems
- Certification and registration of secondary resources are introduced including documentation of technical performance and purity of the materials
- There is a professional demand and market for the major fractions and types of secondary materials, such as reused brick and tiles, aggregates for substituting natural raw materials, timber and construction steel
- Reusable concrete elements are brought on the market

Efficient transformation of CDW to resources presupposes that the market of recycled building materials is implemented and integrated in the general market for building materials. The most important business parameters of recycled materials are logistics related. On-site recycling and short transportation distances are positive parameters. The need for workspace, temporary storage, sufficiently available materials at the right place and time are critical parameters. A very effective and operational resource management system is an absolute condition for the success of recycling CDW and circular economy. The resource management system must comprise a network of resource centres on a local and regional basis with the following tasks, for example:

- Sorting of mixed CDW
- Downsizing and separation of building elements from demolition sites
- Stationary plants and semi-mobile plants for crushing masonry and concrete for unbound materials and recycled aggregates
- Preparing building materials for reuse: doors, windows, timber, steel, glass, etc.
- Stationary and mobile plants for ready-mix RAC
- Fabrication of concrete building materials, pavement stones, road blocks, etc., of RAC
- Storage of materials, building elements, gravel, aggregates, etc.
- Marketplace for recycled materials

7.2.3 Design of new structures

Creative architects always take the opportunities to challenge future designs and exploitation of materials. In addition, utilization of products from demolition of existing buildings and waste materials from the industrial and other sectors have been tested and demonstrated. Some architects are fascinated by the metabolism of buildings and materials in the building and construction sector and waste architecture (see Figure 7.3). However, the building and construction sector is conservative. Any kind of structural functions and changes, production and introduction of new materials require lengthy test procedures for certification and political acceptance.

Figure 7.3 The Upcycled House, Nyborg, Denmark, designed by Anders Lendager, Lendager Architects. The building is made of recycled materials of all kinds. The supporting structure is reused steel containers, floor planks of recycled plastic doors and windows are reused, and so on.

Looking back on the past 50 years, as mentioned earlier in the introduction of this chapter (Section 7.1), no crucial development of the building and construction industry has taken place. Therefore, besides the development of design technologies and some new construction technologies, such as 3D printing, it is estimated that the trends in design of structures will not change significantly in the next 25 years.

Computer and digital information technologies have transformed the construction design methods completely. From the earlier one-dimensional handmade design and manual engineering calculations, the design has changed to computer-aided design supported by Building Information Modelling (BIM), 3D scanning and Virtual Design and Construction (VDC). These trends of high-technology design today will continue. However, looking at the development of smartphones during the past decade, it is impossible to envisage the future developments of the design technologies.

Referring to the Ellen MacArthur Foundation (EMF 2015), the following recommendations for the introduction of the circular economy in the building and construction industry are:

- Applying industrial production processes to reduce waste during construction and renovation, including modular construction of building components or, going even one step further, 3D printing building modules
- Expanding the reuse and high-quality recycling of building components and materials by applying designs for disassembly techniques, material passports, innovative business models, and setting up a reverse logistics ecosystem
- Increasing the utility of existing assets by unleashing the sharing economy (peer-to-peer renting, better urban planning), multi-purposing buildings such as schools, and repurposing buildings through the modular design of interior building components

7.2.3.1 Industrial and modular construction

According to the WEF (WEF 2016), productivity in construction could receive a substantial boost from standardization, modularization and prefabrication. The standardization of components brings many benefits, including a reduction in construction costs, fewer interface and tolerance problems, greater certainty over outcomes, reduced maintenance costs for end-users, and more scope for recycling. Modularization adds to the advantages of standardization, by increasing the possibilities for customization and flexibility, and helping to realize the potential of prefabrication in a factory-like environment. Prefabrication would increase construction efficiency, enable better sequencing in the construction process and reduce weather-related holdups; by such means, it becomes possible to reduce a project's delivery times and construction costs relative to traditional construction methods, and also to create safer working environments.

Modularization has taken place in the brick industry in the previous 2000 years, as described in Chapter 4, Figure 4.18 showing various modules of bricks. The modules are small with an average weight today typically of around 2–3 kg; they are applicable for all kinds of housing, and are timeless and reusable. The concrete industry does also produce modules, such as foundation blocks, lightweight concrete blocks, sewer tubes, and so on. When it comes to structural concrete members, such as columns, beams and slabs no specific standard modules applicable for construction in general are found on the market. Urban development structures are often built by concrete elements designed for the particular project (see Figure 7.4). However, concrete elements are not designed and built for demolition and recycling. When referring to modularization in the construction industry, the *Lego-system* is mentioned as an ambitious model of modules for construction – easy to build and easy to assemble and to disassemble. Concrete modules of this type are seen for low structural purposes (see Figure 7.5). Invention and design of recyclable concrete modules for demolition/disassembly is discussed in the section below.

It is mentioned (WEF 2016) that prefabrication can be applied to a wide variety of project types, ranging from residential housing to large-scale industrial plants. The various systems can be distinguished by their degree of prefabrication: at the simpler end are the mostly two-dimensional building components, such as walls, ceilings or truss elements; then there are modular structures, comprising larger volumetric elements like entire rooms or storeys; and, finally, there are entirely prefabricated assets. Industrial prefabrication of modularized construction elements has been introduced in Denmark in the 1960s and accepted in a number of European countries since the middle of the last century. The success of the industrialized construction technologies is questionable. Due to urban development at the time, buildings constructed by prefab elements did not always meet the social and cultural needs and became slum and had to be demolished, as mentioned above.

Figure 7.4 Typical construction of a high-rise building today; 29 floors, 100 m high, and housing for 500 students. The principle of bearing construction is a central in situ casted core surrounded with concrete elements mounted by a crane. This building concept has been used for more than 50 years. It is expected that this construction principle will be used until 2050 and forward.

The benefit of industrialized construction with respect to waste management and resource efficiency must be subjected to critical evaluation. Industrial production of concrete elements entail a certain production of waste concrete. No exact figures are given. However, up to 10% waste from industrial concrete production is not unusual. Furthermore, the total energy consumption and CO_2 emission from all production processes and logistic activities including extraction of raw materials, industrial production and transport to the construction site must be compared with on-site construction and direct supply of resources. Referring to Section 4.2 'Recycling of Concrete', the benefit of using crushed concrete as aggregate in new concrete is based upon on-site production of concrete.

7.2.3.2 Design for demolition and recycling

Generally, today's structures are designed for a specific purpose in accordance to the requirements of the owner. Besides for temporary structures and buildings designed with a specific purpose and limited lifetime, and special structures such as nuclear power plants and

Figure 7.5 Building modules. Top left: Shows enlarged copies of Lego bricks outside the Lego Company in Billund, Denmark. The Lego system made of durable and reusable plastic bricks inspires ideas for modular construction. Bottom left: Shows the Mijodan block system, Denmark, made of concrete with aggregates of recyclable materials, such as industrial porcelain. Right: Shows the same type of concrete module blocks produced by Theo Pouw, The Netherlands, made of concrete with aggregates of crushed concrete.

offshore structures, nobody gives end-of-life and demolition a thought. However, the design of demolition of high-rise structures from top to ground, as described in Chapter 3, Section 3.7, comprises the discussion of opportunities for dismantling them like a reverse construction. The barriers in dismantling buildings constructed with elements built in late nineteenth century are the separation of the element joints and the lifting of elements. The reinforcement bars of the joints are locked and inseparable, and the joints are grouted with high-strength cement-mortar, which impedes separation of the elements. The lifting hooks and other means for removal of the elements have been cut off, which complicates crane lifting during dismantling of the elements.

Reuse of concrete elements, beams, slaps, columns has never been a serious issue. Some recycling of concrete elements has taken place occasionally. Therefore, there recommendation of the Ellen MacArthur Foundation (EMF 2015) on expanding the reuse and high-quality recycling of building components and materials by applying designs for disassembly techniques is a valuable input for the implementation of the circular economy and to improve resource efficiency in the future.

All structures should be designed with respect to end-of-life and transformation whether considering traditional demolition, renovation or disassembly and reuse of building elements. In particular, special structures with complicated designs, big masses and/or difficult accessibility need to be designed for demolition. For instance, the decommissioning of offshore structures for oil and gas production and wind turbines imply challenges for the demolition of underwater structures and foundations (see Figure 7.6).

7.2.3.3 Design for disassembly

The Danish companies 3XN Architects and MT Højgaard have taken the initiative to challenge the design for disassembly. Referring to *Building a Circular Future* (Jensen and Sommer 2016), the principles and recommendations for design for disassembly are presented. Design for disassembly is defined as a holistic design approach where the intention is to make any given product easy to disassemble into all its individual components. The overall objective is to promote the circular economy and resource efficiency in the building and construction industry. Five principles to consider when designing for disassembly are listed in Table 7.1 (Figure 7.7)

The design for disassembly focuses on the bearing structure and the connection between beams, columns, wall and façades. The connections must be designed and installed according to the principle mentioned in Table 7.1 in such a way that they can carry the designed loads and forces and fulfil the requirements of traditional connections with respect to fire safety. For instance, concrete element connections using steel connectors should be sealed with lime mortar instead of cement mortar.

Figure 7.6 Construction of gravity foundation for a 3,6 MW Siemens wind turbines, Avedøre Holme, on the coast near Copenhagen. (Photo left courtesy of Orsted, Denmark.) Right: Shows one of the foundations of the turbines during construction. The demolition of the heavy reinforced-concrete structures calls for a design for demolition after end-of-life, for an expected 20-year lifetime. Pre-planned demolition for blasting could be a solution.

Table 7.1 Five principles to consider when designing for disassembly according to building a circular future

Materials	Service life	Standards	Connections	Deconstruction
Choose materials with properties that ensure they can be reused	Design the building with the whole lifetime of the building in mind	Design a simple building that fits into a 'large context' system	Choose reversible connections that can tolerate repeated assembly and disassembly	As well as creating a plan for construction, design the building for deconstruction
Use materials of high **quality** that can handle several life cycles	Make the **long-lasting** building elements flexible, so the **short-lasting** elements can be easily changed	Use **modular systems** where elements can be easily replaced	Make the connection **accessible** in order to minimize assembly and disassembly time	Create a **simple plan** for deconstruction, to ensure a quick and easy disassembly process
Use nontoxic materials to provide a **healthy** environment	Make **flexible** building designs that allows the functions to adapt and change in the future	Use **prefabricated** elements for a quicker and more secure assembly and disassembly	Use **mechanical** joints for easy assembly and disassembly without damaging the materials	Make sure that the **stability** of the building is maintained during deconstruction
Use **pure** materials as possible, which can recycled with ease	Think of the building design as a temporary composition of materials and design with the preservation of material value in mind	Create components when the composition of elements becomes too complex to handle	Avoid binders, but if necessary, use binders that are **dissolvable**	**Environment.** Ensure that the deconstruction plan is respectful to the nearby buildings, people and nature

Source: Jensen, K G. and Sommer, J., Building a circular future. Published in 2016 with support from the Danish Environmental Protection Agency.

Recommendations according to Jensen and Sommer (2016):

- In timber connections screws, pins, nuts and bolts should be preferred to nails
- Use common and similar fasteners applicable for a few types of standard tools

- Use easy dissolvable binders instead of glue and sealants
- Use lime mortar instead of cement mortar in brickwork

Considering the lifetime of the building designed for disassembly, the opportunities for recycling are uncertain because the market for recycled materials is unpredictable. The market for building materials is considered as conservative, and it is expected that future markets for reused bricks, timer and steel still exist by end of the lifetime of building constructed today. It is also expected that the market for crushed concrete still exists. Whereas the market for concrete elements is uncertain, as it is today. Therefore, it is a long-term and unsure investment to build concrete structures designed for disassembling and reuse of concrete elements. The building owner has to assess and compare the following two principle options:

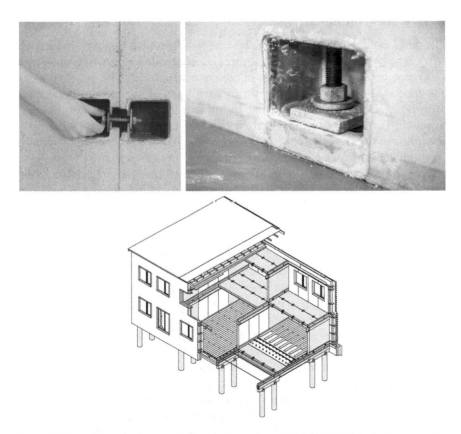

Figure 7.7 Example of concrete element connection designed for disassembly according to the Circle House Project. Bottom figure: Shows a model of the house. (From Jensen, K. G. and Sommer, J., Building a circular future. Published in 2016 with support from the Danish Environmental Protection Agency.)

- Opportunities of future reuse of concrete elements, presupposed design for deconstruction and the total cost, including investment in design of disassembly and value of the elements by the end-of-life of the structure
- Disassembling or demolition of the structure and traditional recycling of concrete, including cost of recycling processes and expected income from the sale of recycled materials

The building owner and his designer must be aware that demolition of structures designed for disassembly requires workers access to the structure and manually work to separate the elements. Attention is drawn to the fact that the trends of demolition work in the future are focusing on mechanical and remote demolition technologies and less handwork. See trends of future demolition earlier in this chapter.

7.2.3.4 3D Printing

The development of 3D printing of concrete structures has taken place for some years. The vision of 3D printed structures is a controlled, rapid, fully automatic and resource-effective production of concrete structures. The industry sector has used 3D print during 30 years for many types of industrial products and prototypes. However, the 3D print technology in the construction industry is new. It is only 3–4 years since the first 3D printed buildings were made according to the Danish Report (3D Printhouse 2018). The technology is based on computer-controlled continuous extruding of mixed concrete, layer by layer, forming a concrete structure in 3D. It is expected that the 3D printing apply to on-site production of concrete structures, as well as industrial production of concrete elements. Asia, China in particular, followed by Europe has been at the forefront of the 3D development and the technology has been proven in full-scale size. Eleven buildings and a number of structures have been 3D-printed in China and other countries. Besides the Danish project presented here, a demonstration project has been carried out in Nantes, France, and two bridges in Germany and Spain (3D Printhouse 2018). In September–December 2017, the first full-scale 3D printing of a building in Europe took place in Copenhagen, Denmark, conducted the company *3D Printhuset* (see Figure 7.8).

The printing was performed with a concrete mix of 0–4 mm sand, 0–8 mm crushed recycled tiles, cement and water in 20 mm high layers. Because of the nozzle dimension, only materials with a size below 8 mm were used. The challenges of the 3D-printing process was to combine timing of print speed, flow of the concrete mix, thickness of printed layer and stiffness of the underlying layer. The success of the printing depends on the optimal match between the printing and the hardening of the concrete layers.

The demonstration project in Copenhagen confirmed the practical performance and the advantage of 3D printing and production of shaped and complicated concrete structures without expensive formwork, which is

Figure 7.8 Three-dimensional (3D) printing of concrete walls of a new building, Copenhagen, November 2017. (Picture of the building design credited 3D Printhouse, Denmark.)

necessary for traditional concrete casting. Besides the challenges of building a large size 3D printer and control of the processes, the demonstration project gained many experiences developing the applicable concrete mixtures. Referring to the project report *3D Printet Byggeri* (3D Printhouse 2018) (in Danish), the sum of the experiences of development of materials for 3D printing were the following:

- The material must be applicable for extruding and pumping. It needs to be rapidly stabilized without formwork and have a suitable setting matching the extruding process
- Cement is applicable for the concrete. High-strength concrete is possible and the concrete can be mixed with a number of additives and types of aggregates. The cement consumption is higher than for normal concrete, which leads to higher CO_2 emission
- Crushed masonry and concrete (particle size below 8 mm) are applicable as aggregate in 3D concrete

- The price of 3D concrete is a little higher than that of normal concrete. However, it is expected that the consumption of concrete would be lower compared with normal concrete

Referring to the WEF (WEF 2016), the development of 3D printing is expected to have a disruptive impact on the construction industry. It is mentioned, that the technology enables the production of purpose-built shapes that cannot be produced by any other method; it promises productivity gains of up to 80% for some applications, together with an important reduction in waste. Construction time for some buildings could shrink from weeks to hours, and customized components could be provided at much lower cost. However, 3D printing in the construction industry is still at an early stage of development. Several issues persist, including resolution problems (large-scale printing often produces rough, chunky results), a trade-off between scale and speed (big printing remains slow, as standard 3D printers are constrained by their size) and high costs. At present, 3D printing is still mostly applicable to low-volume, high-value parts. It remains to be seen how quickly companies will overcome its main technological challenges, and whether they will be able to bring down costs and achieve economies of scale.

3D-printed concrete should be considered as an additional technology for special and complex-shaped concrete structures. The expected disruptive impact on the construction industry, as mentioned by the WEF (WEF 2016), is questionable and hardly realistic for some time to come. With respect to future management of concrete waste and recycling, it is not expected that 3D printing will play a role of importance. However, it should be mentioned that the 0–4 mm fines from crushing concrete could be used for 3D-concrete printing, substituting for natural sand materials.

7.2.4 Construction of new structures

Concrete, masonry, steel and wood will still be the major materials in the building and construction industry. The engineering design of the bearing structures is not expected to undergo significant changes. Façade constructions and materials, insulation and technical installations will undergo current development, such as wireless electricity, digital controlled heating, ventilation and water systems and so. It is expected that construction processes will be improved with respect to effectiveness, speed, health and safety and economy. In addition, improved resource efficiency will reduce the consumption of resources and the generation of waste by recycling of building materials and reduction of waste.

7.2.4.1 Use of recycled materials

Today, the EU, some member states and other states intend to implement the circular economy and promote recycling of materials. Some countries have introduced voluntary agreements on use of building materials, such The Netherlands on use of recycled concrete as aggregate in new concrete, as well as the US Federal Highway Association on recycling concrete in road pavements. However, no evidence of specific public compulsory requirements on the use of recycled materials in the construction industry has been seen yet. In the future, it is expected that the use of recycled materials will take place in the construction industry, as a practice in accordance to specific national and federal technical requirements. For instance, such that a certain percentage, for example 20% or 30%, of raw materials in construction of public and private buildings should be recycled materials.

7.2.4.2 Waste reduction

Construction waste originate from all construction processes:

- Surplus of supplies
- Over ordering
- Adaption
- Spill
- Changes of design during construction
- Defect and wrong materials
- Packaging materials

Referring to the Bio report (Bio et al. 2011), current data does not allow a global distinction between demolition waste and construction waste. The amount of construction waste depends on building processes, time of construction, waste and resource responsibility, and so on. Construction waste is usually less mixed, less contaminated and its recovery potential is higher than for demolition waste. Generally, it is estimated that the amount of construction waste generated on a building site, in relation to the total consumption of the building, is of the magnitude of 5%–15%. Referring to a specific example, presented in Table 3.1, 23 kg pr. square meter has been assessed.

During the last decades, the construction industry in the EU has increased its attention on waste management and resource efficiency. It is estimated that the mentioned megatrends (Section 7.1) of improved design and control of construction processes will reduce the consumption of resources and construction waste significantly.

7.2.5 Use of new buildings

According to the Ellen MacArthur Foundation (EMF 2015), increasing the utility of existing assets by facilitating the sharing economy, better urban planning, multi-purposing buildings, and so forth, is expected in the future. In addition, it is estimated the future use of new buildings will be planned and controlled in such a way that the waste production and need for resources will be reduced. From a resource-efficient point of view, future buildings will be in use for longer durations and cause less waste. From a development point of use with respect to ever-increasing demands for functions, quality, configuration, facilities, and so on, the life cycle of buildings and structures will change more frequently than now. This will lead to increasing amounts of waste materials to be transformed.

7.2.6 End-of-life and demolition of future old structures

The trends of increased construction activity might be accompanied by increased consumption of resources and production of CDW. However, it is estimated that the improved integrated CDW management will increase recycling and substituting for natural resources, according to the targets and the way ahead described in Section 7.3.

Referring to the description of design for demolition and recycling and design for disassembly, it is expected that the end-of-life of future constructions of new structures will be planned for transformation.

7.3 FUTURE DISASTER WASTE MANAGEMENT

The natural disasters and conflicts over the past years have shown a tendency towards rising magnitude and effect on the societies they affect, as for instance the tsunamis in Indonesia (2004) and in Japan (2011), or hurricanes *Katarina* (2005) and *Irma* (2017) in the United States and the Caribbean. The damages in Syria, since the start of conflicts in 2011, exceed the damages of earlier conflicts since the Balkan conflicts in the 1990s. Across 10 Syrian cities assessed by a World Bank Study (World Bank 2017), 27% of the housing stock has been impacted, with 7% destroyed and 20% partially damaged. Across the eight governorates, about 8% of the housing stock has been destroyed and 23% partially damaged. The damage has been particularly high in the health sector, as medical facilities were specifically targeted. At the city level, an estimated total of 316,649 housing units were exposed to impact, with Aleppo bearing the largest share at 64% of the impacted urban housing, followed by Homs at 16%. According to analyses of debris generation in Aleppo and Homs (Disaster Waste Recovery et al. 2017), 14.9 and 5.3 million tons of debris have accumulated, respectively.

With respect to the future DWM challenges in Syria, as well as other conflicts and disasters in the future, the following measures are expected:

- The need for integrated DWM is recognized and prioritized from the start of the response after conflicts and disasters
- All stakeholders, from early emergency to reconstruction, recognize and handle building waste, especially debris bricks, iron and wood

as resources, which means control of clearance work and disposal/depositing of the materials

- Planning and executing of reconstruction work comprise exploitation of recycled/reused materials

7.4 VISIONS AND TARGETS OF CDW MANAGEMENT

As expressed in the introduction of this book, recycling of building materials is an old business. The basic mindset of recycling is based on occasional use of the most accessible resources, for example during reconstruction of buildings and infrastructure after wars and disasters. Until now, most communities do not find recycling interesting, because of the belief in a never-ending availability of raw materials. Therefore, no planning of the use of recycled resources or adaption of the professional building materials' market has ever taken place. During the past 50 years, recycling of building materials has been very low and based on sub-cultures, individual green thinking and academic playgrounds. In 1993, at the time of the third international RILEM symposium on *Demolition and Reuse of Concrete and Masonry* (Lauritzen 1993), and the construction of three dwelling houses in Denmark with 80% of recycled materials, we expected that the introduction of recycling in the construction industry would be just around the corner. However, no crucial steps towards recycling in the construction industry were taken during the following 25 years.

The European Commission generally finds that there is a need for an efficient market for recycled CDW materials. In general, for most EU countries, landfill and backfill are considered the most beneficial solutions for disposal of CDW. Demolition companies are uncertain regarding the disposal of reusable CDW and prefer safe low-tech solutions. Similarly, there are recyclers of recyclable CDW materials, especially secondary raw materials; however, there are uncertainties regarding the quality, purity and availability of these materials. The markets do not develop economies of scale and the amount of returned recycled materials does not meet the potential demand from construction companies. In some cases, technologies are required to ensure that recycled materials meet all technical, safety and environmental requirements for construction products. In addition, there are also sufficient certification procedures that can certify that recycled material meets all relevant material requirements and their functions.

Today, the general opinion with respect to imperishable raw materials has changed. All agree that the resources must be preserved and the consumption of resources must be reduced. Similarly to the 2030 UN Sustainable Goals, the Paris 2015 Climate Agreement, and the priority on resource efficient use of building materials, it is high time for visions and targets for integrated construction, demolition and DWM.

Recycling resources from CDW depends on quality, purity, availability, and so on, as well as the sector's needs and ability to exploit resources. The assessment of visions, goals and initiatives to promote recycling of CDW is based on the above-mentioned perceptions of megatrends and trends towards future construction, as well as the transformation of the building stock and the associated ongoing resource turnover. The basis for this assessment is supplemented with the experiences gained from previous efforts in this area.

7.4.1 UN sustainable development goals

Among the 17 UN SDG 2030 Goals, the goals no. 11 and 12 address the resources of the construction sector. Table 7.2 lists targets addressing CDW and resources, explicitly the 10 targets of UN SDG 11: *Make cities inclusive, safe, resilient and sustainable*, and the 11 targets of Goal 12: *Ensure sustainable consumption and production patterns*.

CDW and resource-efficient management is an international challenge with respect to the UN SDGs (UNSDG 2030). Because of the current collection of data of the construction industry, the EU, the United States and other industrialized countries have a better background for development and improvement of CDW and resource management. However, the lack of background data and experiences does not justify the neglecting of proper management and recycling of CDW.

Table 7.2 List of targets addressing waste and resources among targets of UN SDG Goals No. 11 and No. 12

UN SDG 11 Make cities inclusive, safe, resilient and sustainable

- By 2030, reduce the adverse per capita environmental impact of cities, including by paying special attention to air quality and municipal and other waste management
- Support positive economic, social and environmental links between urban, peri-urban and rural areas by strengthening national and regional development planning
- By 2020, substantially increase the number of cities and human settlements adopting and implementing integrated policies and plans towards inclusion, resource efficiency, mitigation and adaptation to climate change, resilience to disasters, and develop and implement, in line with the Sendai Framework for Disaster Risk Reduction 2015–2030, holistic disaster risk management at all levels
- Support least-developed countries, including through financial and technical assistance, in building sustainable and resilient buildings utilizing local materials

UN SDG 12 Ensure sustainable consumption and production patterns

- By 2030, achieve the sustainable management and efficient use of natural resources
- By 2030, substantially reduce waste generation through prevention, reduction, recycling and reuse
- Encourage companies, especially large and transnational companies, to adopt sustainable practices and to integrate sustainability information into their reporting cycle
- Promote public procurement practices that are sustainable, in accordance with national policies and priorities

7.4.2 EU Visions and targets on resource efficiency

The *Roadmap to a Resource Efficient Europe* (EC 2011) outlines how we can transform Europe's economy into a sustainable one by 2050. It proposes ways to increase resource productivity and decouple economic growth from resource use and its environmental impact. It illustrates how policies interrelate and build on each other. The vision of CDW management is that CDW should be considered and administrated as a resource, substituting natural, primary resources.

The *European Parliament Resolution of 9 July on Resource Efficiency: Moving Towards Circular Economy* (EU 2015) stresses that by 2050 the EU's use of resources needs to be sustainable and that this requires, inter alia, an absolute reduction in the consumption of resources to sustainable levels, based on reliable measurement of resource consumption throughout the entire supply chain. A number of measures are listed in Table 7.3, which refer to the EU Parliament resolution.

Particular attention must be given to increasing the use of recyclable materials and reducing CDW. Examples include the recycling of metals, concrete and glass and the ability of the producers to obtain replies by recycling their products. Recycling of CDW creates job growth. It is typical of work that is carried out locally, which will create job opportunities throughout Europe.

As a basis for the EU's overall resource policy, the Raw Materials Initiative (EU 2008) recommends a three-pillar raw material strategy:

1. Secure access to raw materials
2. Appropriate framework conditions
3. Reduction of EU consumption of raw materials

Despite the fact that the EU's raw material initiative focuses on critical raw materials, including rare earths, the initiative generally aims at reducing the consumption of raw materials by implementing resource efficiency and increasing recovery of secondary raw materials, especially from pure CDW materials. The initiative was followed by initiatives under the above-mentioned pillars:

* Increased resource efficiency
* Use of waste as secondary raw materials
* International management of waste streams

7.5 VISIONS AND TARGETS OF DISASTER WASTE MANAGEMENT

Today UNDP, UNOCH and other UN organizations and international NGOs have taken initiative to establish frameworks and plans for debris

Table 7.3 Overview of vision EU2050, selected targets and measures with respect to CDW (EU 2011)

Vision EU2050

- The EU's use of economy respects resource constraints and global conversion
- All resources – raw materials, energy, air, water and soil – are managed in a sustainable manner

Targets

- By 2050 EU's use of resources needs to be sustainable
- By 2030 increase resource efficiency at EU level by 30% compared with 2014 level
- Promote the use of resource efficiency indicators

Measures

- Full implementation of the circular economy principles and requirements in the building sector to further development the policy framework on resource efficiency in buildings
- Introduction of best available technologies (BAT) and development of material passport based on the whole life cycle of a building
- Considering that, as 90% of the 2050 built environment already exists, special requirements and incentives should be set for the renovation sector in order to improve the energy footprints of buildings by 2050 and to develop long-term strategy for the renovation of existing buildings
- Improvement of recycling through the development of infrastructure for selective collection and recycling in the construction industry
- Introduction of pre-demolition audits (which is an assessment of a building before deconstruction with respect to recycling opportunities) and on-site sorting of recyclable materials
- Increased recycling of concrete
- Development of markets for high-quality secondary raw materials and the development of business based on the reuse of secondary raw materials
- Development of long-term and predictable policy framework to stimulate the
- Level of investment and action needed to fully develop markets for greener technologies and promote sustainable business solutions
- Promotion of industrial symbiosis programmes that support industrial synergies for reuse and recycling and that help companies – particularly SMEs – discover how their energy, waste and by-products can serve as resources for others

Source: European Commission, Directive of the European Parliament and of the Council amending Directive 2008/98/EC on waste, 2015.

management. The Sendai Framework for Disaster Reduction 2015–2030[4] comprises seven goals, including goals to (UNISDR 2015):

- Reduce direct disaster economic loss in relation to global gross domestic product (GDP) by 2030; and
- Substantially reduce disaster damage to critical infrastructure and disruption of basic services, among them health and educational facilities, including through developing their resilience by 2030.

---

[4] The Sendai Framework is a 15-year, voluntary, non-binding agreement which recognizes that the State has the primary role to reduce disaster risk but that responsibility should be shared with other stakeholders including local government, the private sector and other stakeholders (UNISDR 2015).

In addition, the Sendai Framework comprises four priority areas, of which Priority 4 is relevant. Priority 4: Enhancing disaster preparedness for effective response and to 'Build Back Better' in recovery, rehabilitation and reconstruction could form the basis for focus on resources and managing disaster waste materials.

However, no evidence of specific plans for recycling, or forward-looking initiatives to promote recycling and resource-efficient debris management and reconstruction after disasters has been seen in existing guidelines for disaster response.

Referring to the experiences from debris management presented in Chapter 6, the following future scenarios must be considered:

- Continued laissez fair – doing nothing
- General attention on debris management and recycling CDW – doing something
- Deliberate action to promote debris management and recycling CDW for reconstruction – trial and error
- Acceptance, ownership and commitment – full integration of resource management, waste is resources

Today, the general scenario on emergency response features a preference for the laissez fair scenario. Most Red Cross, NGOs and other organizations involved in emergency response remove and dispose debris unplanned and randomly. They use local resources for debris removal without any special care for resources or environmental impact. The steps towards more attention and doing something are very short and easy, while deliberate actions towards recycling for reconstruction are more complicated. Referring to the proposed strategy for integrated DWM in Section 6.4, a list of proposed visions, targets and measures are listed in Table 7.4.

7.6 COMMON VISIONS AND TARGETS

The common visions and targets on saving resources, reduction of waste, resource efficiency and reduction of greenhouse gases must be directed to the waste sector, as well as the construction sector and the humanitarian aid sector. The visions and targets must be relevant and understandable for all stakeholders of demolition, construction and reconstruction. From a waste perspective, it is desired that all building waste materials from demolition, construction and disasters must be managed as resources. Knowing the fact that there will always be a small fraction of hazardous and non-recyclable waste, we are in principle talking about 100% recycling. This is not the case for disaster waste because much of this waste will never be handled. It is desired that resources for construction and reconstruction be managed efficiently; design and work on-site must aim on expedient consumption

Table 7.4 Proposed vision, targets and measures for integrated DWM

Vision 2050
- Waste building materials from disasters and conflicts are handled as resources applicable for recycling and use in reconstruction of buildings and infrastructures
- Reconstruction of disasters and conflicts includes sustainable use of recycled materials

Targets 2030
- By 2030 DWM is prioritized and deliberate in planning of emergency response and disaster risk reduction
- By 2030 the need for integrated DWM is recognized and prioritized from the start of response after conflicts and disaster

Measures
- All stakeholders, from early emergency to reconstruction, recognize and handle building waste, especially debris bricks, iron and wood as resources, which means control of clearance work and disposal/depositing of the materials
- Planning and executing of reconstruction work comprise exploitation of recycled/ reused materials
- Clearance of destroyed buildings and handling of building waste materials take place with regard to sorting materials in fractions
- Opportunities for recycling of debris and other materials and use for reconstruction are recognized and accepted
- Development of recycling scenarios with respect to minimization of transport cost
- Controlled of handling of all hazardous materials, such as asbestos, chemicals, batteries, etc.
- Control of transport and depositing materials during clearance operations
- Planning of sites for collection and recycling of disaster waste materials
- Introduction of best available technologies (BAT) for clearance and recycling operations
- Introduction of damage assessment methodologies with respect to recycling potential of the damaged buildings and structures
- Understanding of various opportunities for highest quality recycling of debris
- Promotion of local employment for clearance and recycling materials on *work-for-cash* principles
- Development of long-term and predictable policy framework for international support to DWM

of resources, while substituting primary natural resources with secondary recycling materials and minimizing the waste of resources. The share of substitution of primary resources with secondary resources depends on the design, requirements, logistics, accessibility to primary and secondary resources, and so on, which need a detailed assessment of the business model (see Figure 5.1). In industrialized communities, realistic goals for substitution of up to 20%–30% of materials might be realistic. We might see opportunities for higher substitution percentages in the reconstruction of houses after disasters and conflicts because of the lack of resources and difficult logistics.

The initiatives and policies in our time of waste prevention, reduction and recycling are raised by the waste sector urged by environmental causes.

Table 7.5 Proposed common vision, targets and measures for integrated demolition, construction and DWM

Vision 2050

- All building demolition, construction and disaster waste materials are managed efficiently and sustainably as resources applicable for substituting primary resources
- Full implementation of sustainability and resource efficiency in design, planning and performance of buildings and structures
- Full implementation recycling of waste materials and sustainable use of recycled materials for reconstruction in response after disasters and conflicts

Targets 2030

- The market for recycled resources is established and accepted
- Circular economy and green change are implemented in the construction sector.
- Building and structures are designed for transformation
- UN organizations, NGOs, donors and other organizations adopt policies on integrated DWM; DWM is prioritized and deliberate in planning of emergency response and disaster risk reduction
- Private companies, especially large and transnational companies, adopt policies on integrated CDW and resource management
- Public procurement practices are implemented, in accordance with national policies and priorities on integrated CDW and resource management
- Increase resource efficiency at EU level by 30% compared with 2014 level
- Recycling of CDW above 90% of the generated CDW
- Substitution of primary raw materials above 10% of the consumption of raw materials

Measures/Actions

Waste-related measures/actions:
- Identification and control of CDW streams
- Introduction of pre-demolition assessment, including assessment of potential recycling and reuse
- Increased recycling of concrete on highest possible level of quality
- Development of technologies, methodologies and standards for decontamination and purity of waste materials
- Requirements for on-site sorting of waste, selective demolition

Construction-related measures/actions:
- Implementation of the circular economy principles and requirements in the building sector to further develop the policy framework on resource efficiency in buildings
- Development of policy framework to stimulate the level of investment and action needed to fully develop markets for greener technologies and promote sustainable business solutions
- Development of markets for high-quality secondary raw materials and the development of business models based on the reuse of secondary raw materials
- Development of guidelines, technical standards and/or certification for use of recycled materials
- Development of LCA-methodologies, technologies and business models and design for transformation of buildings, including design for disassembly and recycling of materials

Response and reconstruction-related measures/actions:
- All stakeholders from early emergency to reconstruction recognize and respect the opportunities for recycling and use of recycled materials
- Development of policy framework for national, regional and community based DWM, including promotion of local employment for clearance and recycling materials on *work-for-cash* principles

(Continued)

Table 7.5 (Continued) Proposed common vision, targets and measures for integrated demolition, construction and DWM

- Control of waste streams during damage assessment, clearance work and disposal/depositing of the materials, including control of all hazardous materials, such as asbestos, chemicals, batteries, etc.
- Development of guidelines for clearance of destroyed buildings and handling of building waste materials, including guidelines for recycling scenarios and business models with respect to minimization of transport cost
- Development of guidelines and technical standards for recycling, recycled materials and design of reconstruction of buildings using recycled materials

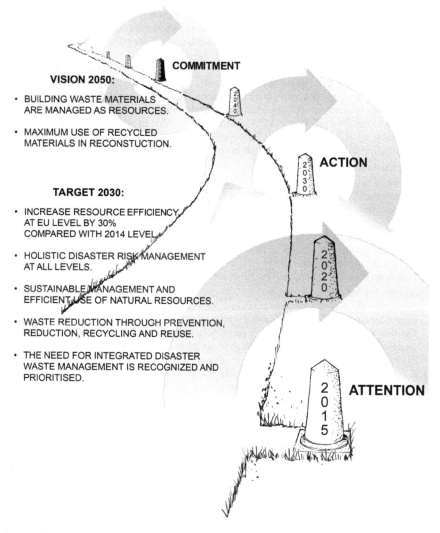

Figure 7.9 Road map of integrated demolition, construction and DWM.

Recycling of waste makes only sense if somebody needs the recycled materials (see discussion in Chapter 2). Therefore, it has been necessary to involve the construction sector, a major consumer of recycled resources. Destruction of buildings caused by disasters and conflicts produces huge amount of building waste materials. Therefore, the development of recycling and substituting primary resources with secondary resources contributes to the response of disasters and conflicts, in particular in response scenarios with shortages of resources.

Based on the UN SDGs and EU visions, targets and measures – together with the experiences and ideas presented in this book, a set of common visions, targets and measures for integrated DWM are presented in Table 7.5. Figure 7.9 shows an illustrative road map of the vision and targets.

Reaching the common visions and targets presuppose dialogue and knowledge sharing, as well as concerted actions provided by the three mentioned sectors: waste, construction and disaster response (reconstruction).

Annex I: Glossary

Aggregate: Sand, stone and similar granular materials which are used in the production of concrete.

Circular economy: Generic term representing the overall vision on preservation of the resources of the planet by recirculating resources and products.

City Concept: Concept/model of a holistic way of thinking and a specific way of doing with respect to the management of buildings and building materials in urban renewal.

Construction and demolition waste: Waste from new building and construction work, or renovation and demolition of buildings and structures (CDW).

Debris: A mixture of building waste, mainly concrete and masonry typically arising from damaged buildings and their demolition, also called rubble.

Demolition: Working process, which serves the purpose of breaking up structures, structural components and building materials.

Disaster waste: Waste generated by the impact of a disaster, both as a direct effect of the disaster as well as in the post-disaster phase as a result of poor waste management.

Disaster waste management: Planning, sorting, collection, handling, transportation, recycling, reuse and disposal.

Dismantling: Working process which serves the purpose of separating and removing intact structural and other building components.

Disposal site:

1. Site for disposal of non-hazardous solid wastes, dumped, compacted and covered daily.
2. Controlled site for hazardous wastes, selected and designed so that leakage of harmful substances is reduced as much as possible.

Dump: Uncontrolled disposal site.

Fragmentation: Working process, which serves the purpose of breaking up concrete, masonry and asphalt to geometrical sizes smaller than the feed opening of the crusher plant.

Further use: Use of materials and products from new building, or renovation and demolition of buildings and structures, for other purposes than the original.

Hazardous waste: Waste that has physical, chemical, or biological characteristics such that it requires special handling and disposal procedures to avoid risk to health, adverse environmental effects or both. 'Hazardous' relates to the situation and circumstances as well as the properties of waste materials. Typical characteristics include: oxidizing, explosive, flammable, irritant, corrosive, toxic, ecotoxic, carcinogenic, infectious, toxic for reproduction and/or mutagenic.

Implosion: Turnover of the structure or a vertical progressive collapse by explosives.

Landfill site: Waste disposal site at which only non-polluted soil and CDW wastes are permitted to be dumped.

Life cycle: The life history of a product from the winning of raw materials to removal as waste, in popular terms 'from craddle to grave'. Life cycle is not a proper term because it is not a true cycle.

Macro-recycling: Mechanical demolition, transport of materials to the nearest recycling facility, crushed by a mobile/semi-mobile crusher (capacity up to 250 t/hour). The recycled materials should be sold substituting natural materials to a price compatible with the price of the natural materials.

Micro-recycling: Manual demolition, sorting, and crushing of debris on-site by the owner/neighbors using mini-crushers or micro-crushers. Reuse of the crushed materials takes place on-site or in the community/neighborhood.

Mini Blasting: Blasting demolition carried out by use of very small explosive charges 5–50 grams and complete protection.

Multiple-building transformation: Change of several buildings in an Old City to a New City.

New building: Transformed old building.

New City: Transformed Old City.

Old building: Existing building to be transformed to new building.

Old City: Existing buildings or building areas to be transformed to a New City.

Partial demolition: Demolition of smaller or larger parts of buildings or structures. Partial demolition is usually part of a preliminary project for renovation and repair work.

Potential amounts of waste: Amounts of waste assessed on the basis of information from authorities, builders, contractors and trade organizations.

Primary raw materials: Virgin or natural raw materials such as ore, wood, sand and gravel.

Recovery: Gaining resources out of waste, e.g. scrap metal.

Recycling: Recycling is the collective term for reuse, recycling, recovery, retrieval and continued utilization, which has the purpose of reducing the amount of waste that is burned or dumped, and to save resources.

Removal: Clearance of destroyed buildings and delivery of debris and other waste to recycling or disposal facilities without subsequent positive utilization.

Renovation: Activities which imply intervention in, or removal of, parts of existing buildings or structures, and which are not included under the concept of demolition. Renovation also includes maintenance activities, which generate waste to the extent that it is of importance in the planning of recycling activities.

Resources: Natural raw materials, water and energy.

Retrieval: Use of materials and products from new building and from renovation and demolition of buildings and structures, after processing but not necessarily used for the original purpose.

Reuse: Use of debris/waste or surplus materials from new building, or from renovation and demolition of buildings and structures after processing, and used for the original purpose and in more or less the original shape.

Rubble: A mixture of building waste, mainly concrete and masonry typically arising from damaged buildings and their demolition, also called debris.

Secondary raw materials: Recycled, recovered or reused materials from construction and demolition processes with technical properties enabling to substitute primary raw materials.

Selective demolition: Method of demolition which aims at the systematic sorting, removal and storage of various fractions of waste materials.

Single-building transformation: Change of a building to reused or new building. Transformation from old life cycle to a new life cycle.

Staging site: Area for temporary stockpiling of debris.

Total demolition: Total demolition involves complete removal of buildings or structures.

Transformation: Processes with respect to building, structures and materials changes from one life cycle to another life cycle and from waste to resources.

Waste: Any material which is generated as a residual product or a waste product from a given activity and which is not reused in the same activity. Waste is any substance which is discarded after primary use, or it is worthless, defective and of no use.

Waste deposit: Waste deposit facility. Distinction is made between landfill site, controlled disposal site, temporary disposal site and special disposal site.

Annex II: Acronyms and abbreviations

ACI	American Concrete Institute
ACM	Asbestos-containing material
ALARP	As low as reasonably practicable
ASTM	American Society for Testing and Materials
BATNEEC	Best Available Technology Not Entailing Excessive Costs
BCD	Beirut Central District
BCPR	Bureau for Crisis Prevention and Recovery of UNDP
BIM	Building Information Modelling
BREEAM	British Research Establishment Environmental Assessment Method
BWMS	Building Waste Management System
CDM	Construction and demolition materials
CDW	Construction and demolition waste
CEN	European Committee for Standardization
CNE	Compagnie Nationale d'Equipment
COP 21	UN Climate Agreement, Paris 2015
3D	Three dimension (e.g. 3D printing)
DAKOFA	Danish Competence Centre on Waste and Resources
DEPA	Danish Environmental Protection Agency
DGNB	Die Deutsche Gesellschaft für Nachhaltiges Bauen – DGNB, German Sustainable Building Council
DMWG	Debris Management Working Group
DoW	Design out waste
DRR	Disaster risk reduction
DWM	Disaster waste management
DWR	Disaster Waste Recovery (UK NGO)
EAHR	Emergency Architecture and Human Rights
EC	European Commission
EIA	Environment impact assessment
EOL	End-of-life
EN	European Standards
EMF	Ellen MacArthur Foundation

EOD	Explosive ordnance disposal
EOW	End-of-waste
EPA	Environmental Protection Agency (USA)
EU	European Union
EUAM	European Administration of Mostar
EWB DK	Engineers without Borders Denmark
FHWA	US Department of Transportation's Federal Highway Administration
GHG	Greenhouse gases
GIS	Geographic information system
GPS	Global positioning system
HABITAT	UN HABITAT
IDP	Internally Displaced Persons
IETC	International Environmental Technology Centre
IHRC	Interim Haiti Recovery Commission
ILO/BIT	UN International Labour Organization
IOM	International Organization for Migration
IRMA	Integrated Decontamination and Rehabilitation of Buildings, Structures and Materials in Urban Renewal 2003–2006
IRC	International Red Cross
ISO	International Organization for Standardization
ISWA	International Solid Waste Orgainzation
JEU	Joint UNEP/OCHA Environment Unit
JPDR	Japan Power Demonstration Reactor
JRC	EU Joint Research Centre
KFOR	Kosovo Force
LBP	Lead-based paint
LCA	Life Cycle Assessment
LCI	Life Cycle Impact
LCIA	Life Cycle Impact Assessment
LEED	Leadership in Energy and Environmental Design, US Green Building Council
LT	Long Term
MDG	Millennium Development Goal
MINUSTAH	United Nations Stabilisation Mission in Haiti
MTPTC	Ministry of Public Works, Transport and Communication Ministère des Travaux Publics, des Transport et de la Communication
MSF	Médecins Sans Frontières (Doctors Without Borders)
MSW	Municipal solid waste
NAC	Natural Aggregate Concrete
NATO	North Atlantic Treaty Organization
NBC	Nahr el-Bared Camp, Tripoli, Lebanon
NGO	Non-government organization
NRA	National Reconstruction Agency (Nepal)

NRMCA	National Ready Mixed Concrete Association (USA)
OCHA UN	Office for the Coordination of Humanitarian Aid
ODA	Olympic Delivery Authority (OL 2012, London)
OECD	Organiszation for Economic Co-operation and Development
PA	Public assistance
PAH	Poly-aromatic hydrocarbon
PARDN	Action Plan for National Recovery and Development of Haiti
PCB	Polychlorinated biphenyl
3Rs	Reuse, recycling and recovery
4Rs	Reduction, reuse, recycling and recovery
RAP	Reclaimed Asphalt Pavement
RCA	Recycled Concrete Aggregate
RAC	Recycled Aggregate Concrete
R&D	Research & Development
REWARD	Recycling Waste Research and Development, 1990–1992
RFP	Request for Proposals
RF	Recovery Framework
RILEM	Réunion Internationale des Laboratoires d´Essaius et de Recherches sur Matériaux et les Construction/International Union on Testing and Research Laboratories for Materials and Structures
SDG	UN Sustainable Development Goals 2030
SEEUR	Services d´Entretien des Equipments Urbains et Ruraux
SoW	Scope of Work
ST	Short Term
STEP	Science and Technology for Environmental Protection (1989–1992)
SW	Solid Waste
SWM	Solid waste management
SWMTSC	Solid Waste Management Technology Support Centre, Nepal
TC	Technical Committee
TLF	Trutier Landfill
ToR	Terms of Reference
UNDAC	United Nations Disaster Assessment and Coordination team
UNDP	United Nations Development Programme
UNEP	United Nations Environment Program
UNEP IETC	United Nations Environment Program – International Environmental Technology Centre
UNESCO	United Nations Education, Scientific and Cultural Organization
UNICEF	United Nations Children's Fund
UNISDR	UN Office for Disaster Risk Reduction
UNMIK	UN Mission in Kosovo
UNOPS	UN Operation Support

UNRWA	United Nations Relief and Works Agency for Palestine Refugees in the Near East
USACE	US Army Corps of Engineers
USAID	US International Aid Department
US FEMA	US Federal Emergency Management Agency
USWMS	Urban Solid Waste Management System
UXO	Unexploded ordnance
VDC	Virtual Design and Construction
VOS	Volatile Organic Substances
WASH	Water, Sanitation and Hygiene
WDR	Waste Demolition Recycling (organization established in Kosovo by the Danish Ministry of Foreign Affairs)
WEF	World Economic Forum
WHO	World Health Organization

Annex III: References

3XN Architects information on renovation project in Sidney and information on project Circle House, mails dated 6th and 22nd February 2018.

3D Printet Byggeri, Afslutningsrapport februar 2018 (Danish), Partnerskabet 3D Printet Byggeri, 2018 (in Danish).

3D Printhuset, The construction of Europe's first 3D printed building has begun. Press release 22 September 2017.

3D Printhuset, visit 11 September 2017 and 15 February 2018 and meetings with Mr. Jacob Jørgensen, 3D Printhuset.

Alaejos, P., Sanchez de Juan, M., Rueda, J., Drummond, R., and Valero, I. Quality assurance of recycled aggregates, Progress of Recycling in the Built Environment. Final report of the RILEM Technical Committee 217-PRE, 2012. Published by Springer 2013.

American Concrete Institute (ACI), Removal and Reuse of Hardened Concrete, ACI International 2001.

American Concrete Institute (ACI), Recycling Concrete and Other Materials for Sustainable Development, ACI International 2004.

Applied Technology Council (ACT) 20, Procedures for Post-Earthquake Safety Evaluation of Buildings, and the ATC-20-2, Addendum to the ATC-20 Post-earthquake Building Safety Evaluation Procedures.

Bio Intelligence Service, Acadis, Institute for European Environmental Policy, Service contract on management of construction and demolition waste – SR1, Final report Task 2, February 2011.

Brundtland, G. H., Report of the World Commission on Environment and Development: Our Common Future, 1987.

California Department of Transportation (Caltrans) information received on Danish Road Directorate visit to Caltrans November 2015.

Carris, J., Demolition waste management on the Olympic Park, Learning legacy, Olympic Delivery Authority, Lessons learned from the London 2012 Games construction project, 2014.

CEMEX, CEMEX entwicklet Beton mit rezykliert Gesteinskörnunbg für Hochbau, CEMEX press info 2017.

CEN/TC 350 Sustainability of construction works.

Copenhagen, Municipality of Copenhagen, Genbrug af mursten, January 2016 (in Danish).

Copenhagen, Municipality of Copenhagen, Genbrug af mursten. Erfaringer fra nedrivning af bygning 16 på Bispebjerg Hospital og genbrug af mursten til nybyggeri på Katrinedal Skole, Vanløse, 2017 (in Danish).

Danish Competence Centre on Waste and Resources (DAKOFA) "Viden om" Den gode kortlægningsrapport 1917 (in Danish).

Danish Competence Committee for Waste and Resources (DAKOFA), Øget kvalitet I genanvendelse og genbrug af bygge-og anlægsaffald i EU, 2015 (in Danish).

Danish Concrete Association recommendations for reused materials in concrete in passive environmental class. DBF publication no. 34, 1989 and Annex 1995 (in Danish).

Danish Environmental Protection Agency, LCA af genbrug af mursten, Miljøprojekt nr. 1512, 2013 (in Danish).

Deloitte, Bio, BRE, RPS, ICEDD, VTT, and FCT, Resource efficient use of mixed wastes – Task 2 – Case study: Construction works in the preparation of the Olympics Games in London 2012.

DEMEX 1998 Strategic study and cost-benefit assessment for demolition and recycling in Ballymun Regeneration, report to Ballymun Regeneration Limit, 30 December 1998.

DEMEX A/S, Building waste management, Post war clearance 1999–2003, territory of Kosovo Exit prepared for the Danish Ministry of Foreign Affairs, DANIDA, 2004, unpublished.

Demolition and Recycling International Buyers' Guide 2017.

De Pauw, C. and Lauritzen, E. K., Disaster planning structural assessment, demolition and recycling. Report of Task Force 2 of RILEM Technical Committee 121-DRG, Guidelines for Demolition and Reuse of Concrete and Masonry. E. & F. N. Spon, 1994.

Disaster Waste Management 2010. Technical report on the quality of debris in Port-au-Prince, Haiti, August 2010.

Disaster Waste Recovery, Roedown Research R² and Urban Resilience Platform, Aleppo: Debris and environmental management for recovery, Phase 1 report Version 1.5, 10th February 2017.

Ecorys and Copenhagen Resource Institute. Resource efficiency in the building sector, 23 May 2014.

Ellen MacArthur Foundation (EMF). Towards the circular economy: Accelerating the scale-up across global supply chains, 2014.

Ellen MacArthur Foundation (EMF). Delivering the circular economy – A toolkit for policymakers, 2015.

Ellen MacArthur Foundation (EMF) Stiftungsfonds für Umweltökonomie und Nachhaltigkeit (SUN) and McKinsey Center for Business and Environment, Growth Within: A Circular Economy for a Competitive Europa, July 2015.

Ellen MacArthur Foundation (EMF), Towards a circular economy: Business Rationale for an accelerated transition, November 2015.

Engineers without Borders (EWB) and Emergency Architecture and Human Rights (EAHR), Appraisal mission, Nepal, capacity building on debris management and recycling of building materials, March 2016.

Eurogypsum, Increasing Energy Efficiency in Buildings with Internal Insulation: The European Gypsum Industry Solutions 2012.

European Commission COM (2008) 699 Raw Materials Initiative 11 April 2008.

European Commission COM (2011) 21 A Resource-efficiency Europe – Flagship Initiative under the Europe 2020 Strategy, 2011.

European Commission COM (2011) 571 Roadmap for a resource efficient Europe 20 September 2011.

European Commission COM (2014) 445 final, Resource efficiency opportunities in the building sector 1 July 2014.

European Commission Communication COM(2011) 13 final: Thematic Strategy on the Prevention and Recycling of Waste/Commission staff working document 2011.

European Commission Directive of 19 November 2008 on waste and repealing certain Directives 2008/98/EC.

European Commission Directive of 25 June 2002 on the minimum health and safety requirements regarding the exposure of workers to the risks arising from physical agents (vibration) (16th Individual Directive within the meaning of Article 16(1) of Directive 89/391/EEC).

European Commission Directive of 26 February 2014 on public procurement and repealing Directive 2004/18/EC.

European Commission Regulation No. 305/2011 (Construction Products Regulation, or CPR), 9 March 2011.

European Commission Research Centre (JRC) rapport End-of-Waste Criteria, by Luis Delgado, Ana Sofia Catarino, Peter Eder, Don Litten, Zheng Luo, Alejandro Villanueva, EUR 23990 EN-200.

European Commission, Communication, Roadmap to a Resource Efficient Europe Brussels, 20 September 2011 COM (2011) 571 final.

European Commission, Council Directive 96/61/EC of 24 September 1996 concerning integrated pollution prevention and control.

European Commission, Decision of 3 May 2000 on list of waste.

European Commission, Directive 2008/98/EC on waste (Waste Framework Directive).

European Commission, Joint Research Centre (JRC) Institute for Environment and Sustainability: International Reference Life Cycle Data System (ILCD) Handbook, General Guide for Life Cycle Assessment, Detailed Guidance. First edition March 2010. EUR 24708 EN. Luxembourg, 2010.

European Commission, Directive of the European Parliament and of the Council amending directive 2008/98/EC on waste, 2015.

European Commission, Regulation (EU) No 305/2011 of the European Parliament and of the Council of 9 March 2011 laying down harmonized conditions for the marketing of construction products and repealing Council Directive 89/106/EEC.

European omission Study Resource Efficient Use of Mixed Waste, last updated September 2016. Reports to be found on http://ec.europa.eu/environment/waste/studies/mixed_waste.htm.

European Parliament (2014/2208 (INI)) 24 March 2015. Resource efficiency: moving towards a circular economy, 2015.

European Parliament, Regulation (EU) No 305/2011 of the European Parliament and of the Council of 9 March 2011 laying down harmonized conditions for the marketing of construction products and repealing Council Directive 89/106/EEC Text with EEA relevance.

European Standard EN 12620:2013 Aggregates for concrete.

European Standard EN 13108-8 Reclaimed asphalt.

European Standard EN 15804-2012 Sustainability of construction works, Environmental product declarations, Core rules for the product category of construction products.

European Standard EN 15978-2011 Sustainability of construction works – Assessment of environmental performance of buildings – Calculation method, 20111.

European Standard EN 206:2013 Concrete – Specification, performance, production and conformity.

European Waste Catalogue (EWC).

Federal Highway Administration (FHWA), Formal Policy on the Use of Recycled Materials, 2002.

Federal Highway Administration (FHWA) Technical Advisory T 5040.37 Use of Recycled Concrete Pavement as Aggregate in Hydraulic-Cement Concrete Pavement, July 3, 2007.

Federal Highway Administration (FHWA) Recycled Materials Policy, 2006 revised 2010.

Gjødvad, J. F. and Ibsen, M. D. ODIN-WIND: An overview of the decommissioning process for offshore wind turbines, MARE-WINT, New materials and reliability in offshore wind turbine technology, Chapter 22, 2016.

Gjødvad, J. F., Qvist, S. and Lauritzen, E. K., Demolition risk management – Blasting of two high-rise buildings in Copenhagen, Denmark, NIRAS A/S, Denmark. European Federation of Explosive Engineers, International Conference, 2014.

Government of Haiti, Action Plan for National Recovery and Development of Haiti (PARDN), Donor Conference New York 31 March 2010.

Government of Haiti, Commune du Développment Humain Durable – Objectifs du Milléneum du Développment, Gouvernement d´Haïti and les Nations Unies en Haïti, 2004.

Government of Nepal, Earthquake Damaged Building Structures Demolition (Removal) Guideline, 2072, 2015.

Government of Nepal, IOM: Kathmandu valley post-earthquake debris management, Strategic Plan, 2014.

Government of Nepal, National Planning Commission: Post Disaster Needs Assessment, Volume A and Summary, 2015.

Government of Nepal, UNEP and leadership environment and development, NEPAL (LEAD): Disaster waste management – Policy, strategy and action plan, Final Report December 2015.

Guldager Jensen, K. and Sommer, J. Building a Circular future, Published by GXN and MT Hojgaard in 2016 with support from the Danish Environmental Protection Agency.

Hansen, T. C., Recycling of demolished concrete and masonry, RILEM report 6, E. & F. R. N. Spon 1992.

Hendriks, C. F., Pietersen H.S. Sustainable raw materials: Construction and demolition. Report 22.

Hendriks, C. E., Nijkerk, A. A. and van Koppen, A. E. The Building Cycle, Æneas Technical Publishers, 2000.

IHRC – Government of Haiti – Debris Sector Background and Call for Proposals 10 August 2011.

IHRC Debris Management Sector Meeting, Thursday, April 7, 2011.

Integrated Decontamination and Rehabilitation of Buildings, Structures and Materials in Urban Renewal (IRMA), European Fifth Framework Programme Energy, environment and sustainable development Key action 4: City of tomorrow and cultural heritage. Contract no. EVK4-CT-220-00092. Project files and report to the EU Commission 2006.

International Organization for Migration (IOM): Environmental impact assessment of post-disaster debris collection, Disposal and management sites, Final report 2014?

International Solid Waste Management (ISWA) – UNEP: Global waste management outlook, 2015.

ISO 14040 Environmental management – Life cycle assessment – Principles and framework.

Janssen, G. M. T., Ho, H. M. and Put, J. A. L., Global assessment of urban renewal based on sustainable recycling strategies for construction and demolition waste. Progress of recycling in the built environment. Final report of the RILEM Technical Committee 217-PRE, 2012.

Jensen, K. G. and Østergaard, C., Information on transformation of high-rise building in Sidney.

Jensen, K. G. and Sommer, J., Building a circular future. Published in 2016 with support from the Danish Environmental Protection Agency.

Kasai, Y., Demolition and reuse of concrete and masonry, Vol. 1 Demolition methods and practice. Proceedings of the 2nd International RILEM Symposium on Demolition and Reuse of Concrete and Masonry, Tokyo, Japan 1988. Chapman & Hall, 1988.

Kobayashi, Y., Disasters and the problems of wastes – Institutions in Japan and issues raised by the Great Hanshin-Awaji Earthquake. UNEP International Environmental Technology Centre, 1995, Earthquake Waste Symposium, Osaka, 12–13 June 1995.

Lauritzen, E. K., Demolition and reuse of concrete and masonry. Proceedings of the 3rd International RILEM Symposium on Demolition and Reuse of Concrete and Masonry, Odense, Denmark, 1993, E. & F. N. Spon, 1993.

Lauritzen, E. K., Economic and environmental benefits of recycling waste from the construction and demolitions of buildings. UNEP Industry and Environment, April–June 1994.

Lauritzen, E. K. Solving Disaster waste problems and comments. Proceedings Earthquake Waste Symposium, Osaka 12–13 June 1995, UNEP-IETC Technical Publication Series No. 2. 1995.

Lauritzen, E. K., Disaster Waste Management. The ISWA Yearbook, 1996/1997.

Lauritzen, E. K., Emergency construction waste management, Safety Science 30, 1998.

Lauritzen, E. K., Recycling concrete – An overview of challenges and opportunities, ACI International SP-219-1, 2004.

Lauritzen, E. K., Lessons from Lebanon: Rubble removal and explosive ordnance disposal. Journal of ERW and Mine Action (2014).

Lauritzen, E. K. and Hansen, T., Recycling of construction and demolition waste 1986–1995. Danish Environmental ProAgency, Environmental Review no. 6, 1997.

Lauritzen, E. K. and Jacobsen, J. B., Nedrivning og Genanvendelse af bygninger og anlægskonstruktioner, SBI Anvisning nr. 171, 1991 (in Danish).

Lauritzen, E. K. and Petersen, M., Partial demolition by mini-blasting. Concrete International, June 1991.

Lauritzen, E. K. and Schneider, J., The role of blasting techniques in the demolition industry. Paper presented at the European Federation of Explosives Engineers International Conference, 2000.

Lendager, A., Lenager Group, information on the Upcycled House, Nyborg, 2013.

Martins, Isabel, Recycling in Portugal: Overview of CDW, Progress of Recycling in the Built Environment. Final report of the RILEM Technical Committee 217-PRE, 2012. Published by Springer 2013.

Medows, D. L. The Limits to Growth, 1972.

Metha, P. K., Greening of the Concrete Industry for Sustainable Development, Concrete International, July 2002.

Middlebrook, R. F. and Mladjov, R. V., The San Francisco – Oakland Bay Bridge. The Structure Magazine, February 2014.

Molin, C. and Lauritzen, E. K., Blasting of concrete localized cutting in and partial demolition of concrete structures. Building Technology SP Report 1988:09, and RILEM Report 6 Recycling of Demolished Concrete and Masonry, edited by T. C. Hansen, 1992.

Møller, J., Damgaard, A. and Astrup, T., Danish Technical University (DTU) LCA af genbrug af mursten. Miljøstyrelsen, Danish Environment Protection Agency Environment, Miljøprojekt 1512, 2013 (in Danish).

Mueller, A., Recycling in Germany, Overview regarding CDW in Germany, Progress of Recycling in the Built Environment. Final report of the RILEM Technical Committee 217-PRE, 2012. Springer, 2013.

Pacheco-Torgal, F., Vivian, T., Labincha, J. A., Ding, Y. and de Brito, J. Handbook of Recycled Concrete and Demolition Waste, Woodhead Publishing, 2013.

RILEM specifications for concrete with recycled aggregates. Materials and Structures 27, 1994, 557–559.

Saveyn, H., Hjelmar, O. et al. Study on methodological aspects regarding limit values for pollutants in aggregates in the context of the possible development end-of-waste criteria under the EU Waste Framework Directive. European Commission Joint Research Centre Report 2014.

Solid Waste Management Technical Support Centre, UNEP and Leadership for Environment and Development, Nepal, Disaster Waste Management – Policy, Strategy and Action Plan, final report December 2015.

Schulz, R. R. and Hendricks, Ch. F., Recycling of Masonry Rubble, Part 2 of RILEM Report No. 6 Recycling of Demolished Concrete and Masonry, edited by Hansen, T. C., E & FN SPON 1992.

Solid Waste Management Technical Support Centre: Disaster Waste Addressing the Challenges in Nepal. Desk study prepared for UNEP IECT, June 2015.

The European Gypsum Industry Solutions 2014

Thematic Strategy on the Prevention and Recycling of Waste.

Theo Pouw Group, www.theopouw.nl. Company information, visit 2017.

UN Declaration Transforming Our World: The Agenda for Sustainable Development, 25–27 September 2015.

UNEP/OCHA, UN office for the coordination of human affairs, Environmental emergencies section Disaster Waste Management, Guidelines, January 2011, Annex XII Contingency planning, 2011.

UN General Assembly, Transforming Our World: The 2030 Agenda for Sustainable Development Resolution, adopted by the General Assembly on 25 September 2015.

UNDP. Guidance Note Debris Management, 2013.

UNDP, Haitian Experience 2010–2012. Technical Guide for Debris Management, 2013.

UNDP, Note on disaster experiences 2010, unpublished.

UNDP, The Ministry of Public Works, Transport and Communication (Ministère des Travaux Publics, Transports et Communications, MTPTC). Haiti National Strategy, Debris Management, Proposal Prepared, edited by Erik K. Lauritzen, UNDP, October 2011, unpublished.

UNEP and ISWA Global Waste Management Outlook 2015.

UNEP International Environmental Technology Centre, Technical Publication Series No. 2 1995, Earthquake Waste Symposium, Osaka, 12–13 June 1995.

UNRWA, UNRWA Nahr el-Bared Camp Reconstruction. NBC Rubble Removal & Explosive Ordnance Disposal Report, Monitoring Report, December 2009.

UNISDR, UN office for disaster risk reduction, Sendai Framework for Disaster Risk Reduction 2015–2030.

United Nations and Framework Convention on Climate Change, Adoption of the Paris Agreement, COOP 21, December 2015.

United Nations Statistics Division, Glossary of Environment Statistics.

United Nations, Rio Declaration on Environment and Development, United Nations Agenda 21 on Sustainable Growth, 1992.

United Nations, The Millennium Development Goals report 2015.

United States Environmental Protection Agency, Beyond RCRA: Prospects for Waste and Materials Management in the Year 2020, 2003.

US Army Corps of Engineers, Debris Planning Cell, Haiti: Proposed Haiti-Earthquake 2010 Draft Debris Management Plan, 14 February 2010.

US EPA Planning for disaster debris, 2008.

US FEMA Debris management guide – FEMA 325. July 2007.

USAID ECAP. Debris removal situation, context and recommendations, February 23, 2011.

Vanderlev, J. and Angulo, S., Construction and demolition waste recycling in a broader environmental perspective. Progress of Recycling in the Built Environment. Final report of the RILEM Technical Committee 217-PRE, 2012.

Vazquez, E., Progress of Recycling in the Built Environment. Final report of the RILEM Technical Committee 217-PRE, 2012.

Vazquez, E., Recycling in Spain: Overview of CDW, Progress of Recycling in the Built Environment. Final report of the RILEM Technical Committee 217-PRE, 2012.

Vazquez, E., Hendriks, Ch. F. and Janssen, G. M. T. Proceedings of the International RILEM Conference on Use of Recycled Materials in Buildings and Structures. Edited by RILEM Publications, 2004.

Waste. RILEM Publication, Cachan Cedex, France, 2000.

World Bank. Haiti Debris Pilot Project, Strategic Assessment Report, Prepared by Integrity Disaster Consultants LLC (IDC) 17 May 2010.

World Bank Group, The Toll of War, The economic and social consequences of the conflict in Syria, 2017.

World Economic Forum, Shaping the future of construction. A Breakthrough in mind-set and technology, May 2016.

Xiao, J., Li, W., Li, J., Poon, C. S., Fana, Y.-h., Ying, J.-W., Huang X. and Tawana M. M., Recycling in China: An overview of study on recycled aggregate concrete Progress of Recycling in the Built Environment. Progress of Recycling in the Built Environment. Final report of the RILEM Technical Committee 217-PRE, 2012.

Yarwood, J., Rebuilding Mostar, Reconstruction in a War Zone. Liverpool University Press, 1999.

Yokota, M., Seiki, Y. and Ishikawa, H., Experience gained in dismantling of the Japan Power Demonstration Reactor (JPDR). Proceedings of the Third International RILEM Symposium on Demolition and Reuse of Concrete and Masonry, edited by Erik K. Lauritzen, 1993.

Index

Note: Page numbers followed by "*n*" with numbers indicate footnotes.

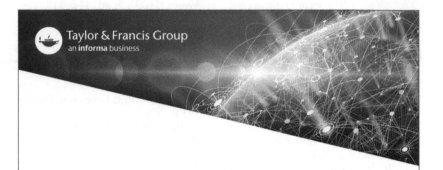